ENVIRONMENT, PLANNING AND LAND USE

T0231598

Environment, Planning and Land Use

Edited by
PHILIP KIVELL
University of Keele

PETER ROBERTS
University of Dundee

GORDON P. WALKER
Staffordshire University

Routledge
Taylor & Francis Group

LONDON AND NEW YORK

First published 1998 by Ashgate Publishing

Reissued 2018 by Routledge
2 Park Square, Milton Park, Abingdon, Oxon, OX14 4RN
711 Third Avenue, New York, NY I 0017, USA

Routledge is an imprint of the Taylor & Francis Group, an informa business

Publisher's Note
The publisher has gone to great lengths to ensure the quality of this reprint but points out that some imperfections in the original copies may be apparent.

Disclaimer
The publisher has made every effort to trace copyright holders and welcomes correspondence from those they have been unable to contact.

A Library of Congress record exists under LC control number: 98070981

ISBN 13: 978-1-138-31308-8 (hbk)
ISBN 13: 978-1-138-31310-1 (pbk)
ISBN 13: 978-0-429-45780-7 (ebk)

Contents

Part Two: Themes and Issues

Part Three: Policy and Management

Figures and Tables

List of contributors

Fergus Anckorn is an environmental scientist with long experience of the coal mining industry in Australasia and the UK. He has worked on environmental aspects of mining with the Australian National University and the New Zealand government. On returning to the UK he was involved with project management of environmental assessments for Wardell Armstrong and now continues this work with Knight Piésold.

Mark Barlow is a research student in the Division of Geography at Staffordshire University. He is undertaking a study of local authority responses to land contamination problems, drawing upon a range of evidence including responses to policy consultation papers.

Susan Buckingham-Hatfield is a Lecturer in Geography and Environmental Issues at Brunel University and Course Director of the MSc in Environmental Change. She is currently researching issues of environmental participation in West London and is particularly interested in how this is gendered. Susan co-edited with Bob Evans 'Environmental Planning and Sustainability' (John Wiley, 1996) and is reviews editor of the journal Local Environment.

Michael Chapman is a Lecturer in the School of Planning and Housing, Heriot -Watt University, He lectures primarily in aspects of European spatial policy. He has researched and written widely on the European dimension of urban and land use planning.

Michael Clark teaches in the Department of Environmental Management at the University of Central Lancashire, Preston. He is a former member of the Council of the TCPA and of its Sustainable Development Working Group.

Nick Coppin is an ecologist and landscape scientist with a broad based experience of environmental assessment, protection, conservation and reclamation. As an acknowledged expert in land restoration and bioengineering techniques, he has written and lectured widely on the subject. He is currently a partner and Head of Environmental Services with Wardell Armstrong, Mining, Minerals, Engineering and Environmental Consultants in Staffordshire.

Evangelos Dimitriadis is Associate Professor of urban planning at Aristotle University of Thessaloniki, Greece. He is a specialist in urban planning and social and historical analysis of urban and regional space, and has written several books, articles and research reports in Greek and in English.

Bob Evans is Head of Geography and Housing at South Bank University, London. He has written and edited several books on environmental planning, is co-editor of the journal Local Environment, and has worked as a town planner in the public, private and voluntary sectors.

István Fodor is Professor of Geography at the Centre for Regional Studies at the Hungarian Academy of Sciences and Head of Department of Environmental Geography at Janus Pannonius University in Pécs, Hungary. He has more than 100 publications and is the editor of the series 'Studies of Environmental Protection'.

Pál Golobics graduated from the University of Economics of Budapest. He currently works at the Department of General Human Geography and Urban Studies at the Janus Pannonius University of Pécs as a senior lecturer. He specialises in three major fields, which are international regional co-operation, cross-border co-operation and the general theoretical issues of international integration.

Sarah Hatfield recently completed a doctoral research project on the use and perception of derelict urban space at the University of Keele. She is currently teaching geography in the West Midlands.

Philip Kivell is Reader in Geography in the Environmental Social Sciences Department at Keele University. He has research interests in a number of fields of applied geography and planning, notably in land use and community health and has published widely, especially concerning land dereliction and restoration.

Christos Th. Kousidonis received his doctoral degree for a dissertation on regional planning he prepared with Professor Lagopoulos. He has been a special lecturer at the Aristotle University of Thessaloniki and is currently in private practice in architecture and town planning.

Alexandros Ph. Lagopoulos is Professor of Urban Planning at Aristotle University of Thessaloniki, Greece, and holds a doctorate in Social Anthropology from the Sorbonne. Co-editor of The City and the Sign (Columbia University Press, 1989) and author of Urbanisme et Semiotique (Anthropos 1995) he has written widely on the theory and the semiotics of space and urban planning.

Tim Marshall is a Lecturer at the School of Planning, Oxford Brookes University. He trained as a planner and worked for several years in practice in Birmingham and London. After a doctorate and research in Spain in 1990-91 he has continued to work on the environmental dimensions of planning and of economic change in Britain, Spain and elsewhere in Europe.

Derek Pratts is a Senior Lecturer in the division of Geography at Staffordshire University, with expertise in areas of water resource management, environmental risks and pollution control policy. His current research is examining the developing framework for air quality management in the UK with reference to case study areas in the West Midlands.

Peter Roberts is Professor of European Strategic Planning at the University of Dundee. His research interests relate to regional and strategic planning in the UK and Europe, the spatial dimension of sustainable development and the theory and practice of urban and regional regeneration. His previous books and publications have included Managing the Metropolis, Environmentally Sustainable Business; a Local and Regional Perspective, Energy Efficiency and Housing, Europe; A Handbook for Local Authorities and Regional Strategy and Partnership in European Programmes.

Marius Schwartz lectures in water management at the Planning Department, Faculty of Spatial Sciences, University of Groningen in the Netherlands. He is a researcher on the planning aspects of water management, especially with regard to the relationship between different planning systems and the organisations involved.

József Tóth graduated from the Jozsef Attila University of Szeged. Currently he is a university professor at the Janus Pannonius University of Pécs, head of the Department of General Human Geography and Urban Studies, and Vice Chancellor of the Janus Pannonius University. He specialises in social and human geography in general, within which his major interests are settlements, population, and economic and social issues.

Gordon Walker is a Principal Lecturer in the Division of Geography at Staffordshire University. His research interests cover a range of risk, energy

Acknowledgements

The chapters in this book originate largely from a conference on Environment, Planning and Land Use held by the Planning and Environment Research Group of the Institute of British Geographers at the University of Keele. The editors would like to thank the Ashgate Publishing Group, The British Academy, Keele University Geography Department (now the Department of Environmental Social Sciences) and the Royal Geographical Society (with the Institute of British Geographers) for financial and other support for that conference. Thanks also go to editorial staff at Ashgate, especially Sarah Markham, Anne Keirby and Pauline Beavers, to Sue Allingham at Keele who typed the manuscript and to Rosemary Duncan at Staffordshire University for preparing the illustrations.

1 Introduction: environment, planning and land use

Philip Kivell, Peter Roberts and Gordon Walker

An overview

Whilst the theme of sustainable development has provided a semblance of unity to the debate on how best to design and implement policies that allow for reconciliation between the economic and environmental objectives of society, in reality, one has only to scratch the surface of this seeming unity in order to discover a multitude of different perspectives and methods of interpretation. However, despite the rich diversity which is evident in the detailed analyses presented in this volume, it is possible to detect a common sense of purpose in many of the contributions. This common sense of purpose is distinguished by the acknowledgement of the important role played by spatial considerations in the analysis of issues and the generation of policy.

Moving beyond the broad rhetoric of sustainable development, a major challenge confronting both academics and practitioners is how best to analyse, design and implement solutions to the environmental, economic, social and political problems that are encountered throughout the regions and nations of Europe. These are problems that are common to both east and west, and learning from individual responses to such problems can help assist in the creation of a pool of knowledge and expertise that will continue to be of considerable assistance in generating a collective response.

This search for explanations and solutions, together with the value which is attached to the exchange of experience, helps to define the primary purpose of this book. It is remarkable how frequently the same or similar problems occur almost irrespective of particular environmental characteristics or political regimes. This coincidence in the occurrence and manifestation of problems allows for the definition and testing of alternative solutions. Whilst it is reasonable to argue that a particular array of problems which is evident at an individual place may be

1

considered to be a unique event, it is equally reasonable to suggest that lessons that are of value in one place should be identified and transferred elsewhere. Learning by doing is important, but costly mistakes can be avoided through the replication of a proven approach. This is not to suggest that blind duplication is the method of diagnosis or prognosis advocated herein, nor is it intended that the dissemination of good practice should deteriorate into a mindless attempt to ape practice elsewhere and thereby create sterile uniformity. Rather, the purpose of this book is to demonstrate the range and the appropriateness of the explanations and solutions on offer, and to suggest that many of these offerings may prove to be valuable as models for the planning and management of the environment.

The contributions to this volume were originally presented as papers at a conference organised by the Planning and Environment Research Group of the Royal Geographical Society (with the Institute of British Geographers) held at Keele University in April 1996. The scope and broad structure of the contributions presented at the conference were defined with the requirements of this book in mind and specific emphasis was placed on six themes. These themes provide coherence and provide a common organising framework for the chapters contained in the present volume. In addition, these themes reflect the breadth of the content of the current discourse on the nature of sustainable development as applied to spatial planning, environmental management and the use and regulation of a range of resources including land.

The first theme has already been mentioned and is a distinguishing feature of the majority of the contributions to the present book. This spatial perspective on planning, environment and land use adds an often ignored fourth dimension to the normal triad of structural, functional and temporal concerns. By adding this fourth dimension, the intention is to provide a total view of the entire environment and planning field of inquiry, rather than duplicating other treatments of the subject, most of which have generally restricted themselves to a single issue or, at best, a three-dimensional view. Most of the chapters incorporate a spatial perspective, and the various explorations of this theme provide evidence of the value of such an approach. We see, for example, the results and consequences of managing water resources through the adoption of an organizational structure that is not entirely coincident in either space or the cycle of programme revision with the land-use planning system. By way of contrast, other examples demonstrate the value of adopting a total approach to the quartet of spatial, structural, functional and temporal concerns in the search for new explanations of socio-political organization or, at a more mundane level, the creation of regional and local frameworks for the implementation of Local Agenda 21.

A second theme that is explored in this book is the considerable influence exerted by socio-political structures over matters of environment and land-use. Society gets the environmental quality that it deserves, or that it is willing to pay for, and this factor frequently governs the ability of planning and environmental management systems to get to grips with what are often deep-rooted, extensive and expensive-to-solve problems. The processes of political and economic choice which govern policy, together with the strength of the resulting legislation, mirror society's attitude towards the environment. In east or west - it appears to make little difference in which sphere of Europe a country is located - a general approach to public policy which favours economic exploitation can overwhelm the environment and thereby render individual efforts at improvement futile. Social learning, to borrow Hajer's phrase (Hajer, 1996), is a pre-requisite for the establishment of a lasting change in attitude, and this change in attitude should exert influence on all sectors and forms of activity.

But how will we know when a lasting change of approach has been successful in bringing about an enhanced environment? The answer to this question provides the substance of the third theme: how to measure and assess actual or proposed changes to the condition of the environment and to the other elements of sustainable development. A number of the contributions explore this theme and provide convincing evidence to support the oft-quoted maxim: if you can't measure it, you can't manage it.

The fourth strand of inquiry, which is evident in a number of the chapters, is the desirability of encouraging the adoption of planning and environmental management systems which encapsulate both a top-down and a bottom-up perspective. People live in places not sectors of activity, and most people reserve their strongest protests for situations when undesirable projects are proposed in their back yard. The culture of the NIMBY is a manifestation of the protective instinct of a society facing a difficult process of choice. This process of choice, if managed with care and sensitivity, can help to empower citizens and allow them to gain greater control over planning for their own immediate environment. This is a message which is equally relevant in Greece and Greater London, and it also yields lessons that are applicable in other policy fields.

A fifth area of exploration which is incorporated in a number of the chapters in this book, relates to the importance of developing methods of analysis and policy systems that are based upon a comprehensive view of all the key elements of planning, environment and land use. Equally important is the need to apply methods of treatment that consider the full range of issues, and which allow for interconnections to be established between different types of administration and between different elements of the socio-economic-environmental system in question.

The final theme discussed in this book reflects the value of transnational and transcultural studies. As is demonstrated by case studies set in both eastern and western Europe, there are many common environmental problems and few of the impacts of these problems are confined to a specific country of origin. The most obvious examples of transfrontier effects relate to the various sources of pollution, and the transmission of the consequences of pollution to adjacent nations. Whilst

3

this is the most visible and, perhaps, the best known example of the transfrontier nature of environmental problems, other less easily detected and more difficult to trace events also occur. For example, western nations have often regarded east-central Europe as a convenient and cheap dumping ground for industrial and domestic waste that cannot be easily or legally disposed of in the west. Equally importantly, the west currently encourages the production of goods in the less harshly regulated economies of the former Soviet bloc. Today, new environmental transnational challenges have emerged as a consequence of economic development and the opening up of central-east Europe to market forces.

These six themes provide a general platform for the contributions made by the following chapters to the analysis of planning, environment and land use issues in a variety of different regions and localities throughout Europe. The scale of concern varies between the chapters: from the micro-level discussion of land use and urban form in a Greek village, to continental-scale assessments of progress with the spatial and environmental policies of the European Union. Equally, the focus of concern differs according to the emphasis placed upon a particular aspect of the subject. Some chapters offer a specific focus on water, on pollution, on derelict land, on urban form or on infrastructure, for example, whilst others take a broader cross-cutting view and emphasise topics such as spatial integration, methods of environmental assessment or the implementation of policy. Whatever the emphasis, the contributions are all marked by a common concern with the desire to apply academic expertise and knowledge to the enhancement of environmental quality.

The structure and organization of this book

The present text is organised in three parts. Whilst each of these parts represents a discrete area of expertise and activity, there is also a high degree of interconnectivity between the parts, with a natural progression from the background theory and context in the first part, through a more detailed discussion of a variety of specific themes and issues, to the development of policy and management agendas in the third part. This progression is deliberate, and it was designed in order to demonstrate both the value of sound analysis that is rooted in detailed research, and the need to encourage the clear assessment of what can be achieved.

Michael Chapman provides the opening contribution, which offers an analysis and assessment of the origins, development and the growing importance of European spatial perspective for the future evolution of planning, environment and land use. Chapman's focus of attention is on the urban scene, and it is at this level that the failings of many previous policy efforts can be observed. In this contribution particular emphasis is placed on the evolution and competence of

4

European Union policy and the problems experienced in attempting to implement this and other associated policies..

The theme of European policy is extended in the second chapter of the first part. Peter Roberts explores the origins and consequences of some of the major environmental problems experienced in both eastern and western Europe. It is argued that the fundamental cause of many of these environmental problems is the dislocation which can be observed between the economic and environmental agendas evident in previous eras of development. The case for a paradigm shift, away from the tonnage ideology of the past and towards the adoption of a development model based on ecological modernization principles, is presented in this chapter. This proposal is illustrated by reference to a number of experiments in the ecological modernization of local and regional economies.

The focus on socio-political issues as a counterbalance to the dominance of environmental and economic concerns, is extended in the following chapter. József Tóth and Pál Golobics trace the origins of the spatial and environmental problems of many of the broader regions of central and eastern Europe, and they argue that geopolitical considerations can be seen to work against the achievement of either environmental progress or economic efficiency.

Tim Marshall's contribution in Chapter five takes a somewhat different tack, albeit related to the first three chapters in terms of its basic philosophy. His chapter examines how the water and energy sectors are being affected within two regional contexts, Catalonia and the English West Midlands, and considers the difficulties generated by the simultaneous emergence of new environmental objectives and drastic sectoral economic change. Attention is also given to the provision of regulation and especially the spatial dimension of regulation through some form of regional planning.

The second part of the book presents a number of specific themes and issues. Diversity of problem and variety of response can be observed in the contributions made to this part.

Nigel Watson provides the first contribution, his selected topic is the integration of land and water management in England. Even though the integrated management of land and water has been identified as a fundamental principle for sustainable development, current practice fails to achieve the required level of integration. A case study is presented of nitrate pollution and of the management response to the call for closer integration.

In contrast to the difficulties encountered in England with regard to the integration of land and water management, the Dutch experience demonstrates the benefits that can be derived from greater co-ordination and integration. In his study of Dutch spatial planning and water management, Maruis Schwartz outlines the evolution of efforts aimed at the closer integration of the two planning systems, discusses the difficulties encountered in achieving coherence and evaluates the costs and benefits associated with such an approach.

5

Gordon Walker, Mark Barlow and Derek Pratts offer a different perspective on the relationship between land and environment. In Chapter eight they examine and evaluate the role of central government and local authorities, and of land use planning in particular, in the management and control of environmental risk. Two case studies are utilised in this chapter, of major industrial accident hazards and of contaminated land risks, in order to illustrate the issues that the planning system is asked to handle and to demonstrate the limits of the capacity of this mode of regulation.

The problems and potentials of derelict land provide further evidence of the management capacities of the public policy system, this time in relation to market conditions and to the community benefits that may emerge on some derelict sites. In their study of derelict sites in North Staffordshire, Philip Kivell and Sarah Hatfield report on survey findings from a series of sites and conclude that such sites can offer both wildlife habitats and a valuable community resource.

In another region of Europe and in relation to a different facet of community life, Evangelos Dimitriadis, Christos Kousidonis and Alexandros-Phaedon Lagoupulos consider the difficulties involved in adapting the spatial form of villages in order better to meet the requirements of sustainable development. This chapter provides an insight into the challenge of micro-scale urban restructuring in order to meet both social and aesthetic requirements, whilst also enhancing economic opportunity. An additional element that is emphasised in the chapter is the importance of securing community participation in the restructuring process.

The final chapter in the second part illustrates the social and economic impediments to the achievement of sustainable development in Hungary. Istvan Fodor's contribution discusses the importance of ensuring the simultaneous progress of economic and environmental agendas. However, as the chapter illustrates, such a coincidence will not occur without intervention, and the Hungarian model is offered as an example of an attempt to manage the processes of change.

In the final part of this book the emphasis switches to questions of policy and management. Four chapters present various aspects of this topic, and they all illustrate the very real difficulties involved in attempting to translate the objectives of policy into solid achievement.

Fergus Anckorn and Nick Coppin examine the role of environmental assessment in the integration of environment and development. Their contribution outlines the origins and operation of environmental assessment, and also illustrate some of the difficulties encountered in applying this method of assessment to mining projects. An important conclusion from this analysis is the desirability of applying environmental assessment to mining policy and plans, rather than leaving matters to the consideration of individual project applications.

A wider view of environmental quality and control is presented by Michael Clark in his examination of quality assurance for planning and environmental management. All too often public objectives are not achieved through either the land use planning system or other public policy instruments. In the case of transportation the results of failure are evident in poor service provision and the environmental consequences of a poorly integrated provision.

Progress with Local Agenda 21 provides the core of the contribution made by Bob Evans. His assessment of the ways in which selected local authorities have responded to the call to complete their Local Agendas, demonstrates the different institutional representations and understandings of the notion of sustainable development. Despite considerable variations in the rate of progress, Chapter fourteen is able to report substantial achievements, especially in the field of environmental regulation.

The final contribution is made by Susan Buckingham-Hatfield. She considers the important, but often neglected, subject of public participation in the formulation of local environmental agendas. This dimension of policy making and application often presents a considerable challenge to local authorities and may force them to rethink the way in which they work with their local communities. Through a case study of a London local authority, she examines the ways in which a local authority has responded to the challenge of engaging the public in the environment debate. The results suggest that there is considerable room for improvement, a message that provides an appropriate finale for a book that reports the results of research in progress.

Reference

Hajer, M. A. (1996), 'Ecological Modernization as Cultural Politics', in Lash, S., Szerszynski, B. and Wynne, B. (eds), *Risk, Environment and Modernity*, Sage: London.

Part One
THEORY AND CONTEXT

2 European spatial planning and the urban environment

Michael Chapman

Introduction

> The expansion of the European Union to the north and east should motivate Brussels to give cities a more prominent place in European policy. Without healthy cities there will be no future for our countries. Dr Abraham Peper, Mayor of Rotterdam (City 2020, p.11).

Although 80 per cent of the population of the European Union (EU) live in urban areas the European Commission (EC) has no direct competence for urban affairs. There is no integrated framework for urban policy and planning at the European level as there is for transport, the environment or regional policy. In keeping with the principle of subsidiarity, as outlined by the agreement of the European Council in Maastricht, December 1991, urban affairs and urban land use policy remains solely the responsibility of local and regional authorities under the overall direction of the individual Member State. While the subsidiarity argument is well rehearsed by some of the member states thereby halting any further encroachment by the Commission into urban matters, there is mounting evidence which supports the claim that Europe is increasingly active at the urban scale and that the initiatives and policies which emanate from the EU have important direct and indirect impacts on the urban environment (Williams, 1996; Chapman, 1995).

The 1990s have witnessed a resurgence of interest in supranational spatial planning at the European level. As Williams (1996) notes, this is partly due to the forces of European integration enhanced by the completion of the Single European Act and reflects wider actions undertaken by the Commission to improved pan European transportation networks and infrastructure like the channel tunnel. While there is still relatively limited experience of spatial planning at the European level, Europe is catching up and learning fast. Spatial planning is perceived to be a key mechanism through which the aims and objectives of the EU can be attained. The move towards a common European spatial development perspective is therefore seen as a necessary platform for constructing European economic, political and social Union. The objectives of economic and social cohesion and the further integration of the EU are difficult to achieve if spatial disparities in wealth, job opportunities and services continue to persist between the member states, the regions and cities of Europe. The purpose of this chapter is to review the significance of the EU in urban policy and land use planning in the city. The chapter goes on to debate the reasons why, and the ways in which, European policies have now become more important for many towns and cities across Europe. As a consequence of the 'Europeanization' of urban policy and urban land use planning, limited consideration has been given to the precise role of the EU in urban matters. While evidence exists to support the case for an urban dimension to be integrated more fully into the European policy process, the Commission has no formal competence for such intervention. However, there is growing political pressure for the EU to address the spatial concentration of economic, environmental and social problems which are present in many European cities. This concern could ultimately bring European policy action into direct conflict with the national governments of the individual member states. The failure to reach a common perspective on the future spatial development of European cities and the role urban authorities play in the policy process could hinder the collective European goal of integration, cooperation and enlargement of the Union. The implications of this gradual inclusion of the urban dimension to European policy are far reaching and go beyond the more rudimentary issues of economic competitiveness and the efficient functioning of a European urban hierarchy. Instead this debate is at the very centre of the future spatial development of an integrated Europe.

Cities at the heart of the European policy agenda?

A key objective of the EU is to address all aspects of economic and social disadvantage at the European, national, regional and local levels. European support for locally based development strategies include both rural and urban areas but in a period of European political and economic integration there is a growing concern that the most acute forms of economic and social disadvantage are to be found in towns and cities. Even in the relatively favourable economic circumstances of the late 1980s there appears to have been a continuation in the economic and social divisions created between the rich and the poor in many urban areas. To explain in more detail the pressures placed upon urban economies Hall (1994) identified seven major forces which have influenced the performance of the European urban hierarchy. These forces are the globalization of the world

12

economy and the formation of continental trading blocs; the transformation of Eastern Europe; the shift towards the informational economy; the impact of transport technology; the impact of information technology; the role of urban promotion and boosterism; and the impact of demographic and social change. He went on to argue that the process of globalization has redefined the comparative advantage of certain cities, simultaneously leading to the de-industrialization of many older cities and the emergence of a very select group of growth orientated 'Eurocities'. Hall (1994) also noted that the transition of countries in central and eastern Europe provides a new stimulus for the concentration of investment in capital cities and second order cities in the European urban hierarchy. This view is supported by Newman and Thornley (1996) who observed that economic change has had contrasting impacts in different parts of Europe. They develop this argument further by stating that it is partly by adaptation and redevelopment of cities that different countries have been able to succeed in international economic competition (Newman and Thornley, 1996, p.11). The importance of cities and urban economies in the future spatial development of Europe is therefore likely to derive from two principal influences. First, the new internationalization of labour and capital and the process of globalization and second, the economic, social and spatial consequences of the Single European Market (SEM) and the commitment towards further economic and monetary integration. The completion of the SEM coupled with the impact of global economic forces has only intensified the competition which already existed between European cities (CEC, 1992a). It is in this context that town planners and urban policy makers struggle to regenerate the most deprived neighbourhoods in the urban economy.

Over the past two decades Europe has experienced rapid and often complex changes in its economic and social situation. While the majority of European citizens have enjoyed the benefits of improved economic and social conditions brought about by a frontier-free Europe, there is a significant and growing minority who have experienced persistent unemployment, poverty and other forms of economic and social exclusion. Terms such as the excluded, the marginalized and the underclass are commonly used to draw attention to new emerging patterns of social and economic inequality in the urban environment (European Foundation for the Improvement of Living and Working Conditions, 1994). Almost all European cities have some areas or districts which have been run down or suffer from urban decay. It is precisely because of the spatial concentration of excluded groups in the urban environment that city authorities are now arguing the case for a stronger role from the Commission. The challenge for the EU in the next five years is now seen in terms of how the needs of a significant and growing minority of European citizens can be integrated into the overall policy framework of achieving economic and social cohesion.

Although there is no explicit European urban policy it is evident that urban areas benefit both directly and indirectly from European policies and programmes. Across Europe, cities and metropolitan areas, receive significant levels of financial support from the EU. This position was further strengthened in 1993 with the introduction of a further review of the implementation of the Community Structural Funds (CEC, 1993a). Since the late 1980s the EC has become increasingly interested in the role that towns and cities play in economic competitiveness, social cohesion and environmental issues. The urban dimension to EU polices was initially recognized in the publication of the 1990 Green Paper on the Urban Environment (CEC, 1990). The Green Paper identified the range of problems being experienced in European cities and suggested that the Community could take certain lines of action in areas such as transport, historical heritage, environmental heritage, industry, energy management, urban waste and urban planning. While the Green Paper was mainly concerned with urban environmental quality it nevertheless represents an important step towards the establishment of a European perspective on urban spatial policy. The Green paper was also the first time that the Commission broached the issue of whether the Community wanted to extend financial support for urban regeneration and environmental improvement over and above previous levels as allowed through the operation of the Structural Funds. Although not acted upon, the Green Paper went a long way in raising the level of awareness of urban issues and did catch the imagination of many politicians, professionals and planners (Williams, 1996).

The importance of understanding the impacts of European wide spatial change was further enhanced by the publication of the report Europe 2000-Outlook for the Development of the Community's Territory, in 1991, (CEC, 1991). This report demonstrated that the Commission's policies had impacted, and would continue to impact, on the land use and physical planning in urban areas of individual member states. As part of the research work towards the publication of the Europe 2000 report the Commission also financed several background papers in a series entitled Regional Development Studies. One such report 'Urbanization and the Functions of Cities in the European Community', (CEC, 1992a) aimed to assess the contribution that cities had made to the changing Europe during recent decades and to identify the broad implications for cities within the European community during the 1990s (CEC, 1992a, p.11). The research work was undertaken in 24 European cities during 1990 and 1991 and the purpose of the study was to understand the dynamics of the economic, social and environmental changes at work in European cities. One observation drawn from this study was that the function of European cities was continually changing and a key factor to explain this change was the importance given to the creation of the Single Market. It was suggested that the most successful cities in Europe would be those which had the ability to compete in the Single Market particularly through diversification of their local economic structures, the attraction of inward investment and the servicing of new markets throughout the whole of the Community. Examples of these dynamic cities included; Hamburg, Rotterdam, Dortmund, Montpelier and Seville. On the other hand, those cities which failed to adapt to competitive pressure would more than likely be cities with severe structural weaknesses in their urban economies, possessing inadequate infrastructure and communications provision and often being associated with branch-plant economies. Cities in this position included Marseilles, Dublin and Naples. In conclusion the study argued that in the search for a more economically dynamic and socially balanced Europe, support for cities should in future be a greater priority for the Community (CEC, 1992a, p.213)

The follow up to Europe 2000 was the publication of Europe 2000+:Cooperation for European Territorial Development, in 1995, (CEC, 1994c). Although the report was completed by 1994 it was only made available for discussion in the spring of 1995 (Williams, 1996, p.104). This report, in accordance with the decisions adopted at the Edinburgh European Council in December 1992, promotes transnational, cross border and inter regional co-operation. The aim of Europe 2000+ was to provide a framework for cooperation over territorial development between planning authorities in the member states, including other countries, in central and eastern European, the former USSR and the southern Mediterranean, and the Commission itself. Cooperation on European spatial planning matters was subsequently supported by the informal meetings of European ministers at Liege in November 1993 and Corfu in June 1994. These meetings stressed the importance of a better understanding of the various forms of spatial planning activities throughout Europe, for member and non member states alike. The urban dimension was discussed in the Europe 2000+ report and it is expected that urban policy and the performance of European cities will also form part of the European Spatial Development Perspective (ESDP) which dates from the informal meetings of spatial planning ministers in Liege in November 1993. The ESDP is intended to set out a common understanding between the Commission and the member states on what exactly a European spatial policy should be achieving. To date little has been said about the precise details and specific aspects of the ESDP, but at the time of writing the first draft is expected shortly.

A more explicit reference to the urban environment can be found in the White Paper on Growth, Competitiveness and Employment (CEC 1993c). This urges the member states to consider how the foundations for achieving sustainable economic development can be established to enable the European economy to withstand international competition while creating much needed employment. The Commission recommends that the EU set itself the objective of creating at least 15 million new jobs, thereby halving the present rate of unemployment by the year 2000. The level of European unemployment was the primary motivator for the publication of the White Paper, although it explicitly acknowledges that no single solution could provide an effective response to the problems of widespread and persistent unemployment. Five priorities for action were identified as the way forward to a new sustainable development model. The parameters of the model emphasise the relationship between growth, competitiveness and employment, better use of environmental resources and the achievement of improvements in the quality of life. The priorities highlighted by the White Paper are as follows:

• Making the most of the single market
• Supporting the development and adaptation of small and medium-sized enterprises
• Pursuing the social dialogue that has, to date, made for fruitful cooperation and joint decision-making by the two sides of industry, thereby assisting the work of the Community
• Creating the major European infrastructure networks
• Preparing forthwith and laying the foundations for the information society (CEC, 1993c).

To achieve these goals special emphasis is laid on the importance of partnership between the public and private sectors. Only in a few instances does the White Paper actually make reference to the role of urban local authorities in this partnership, but it does emphasis the need to achieve an improvement in the quality of life through the promotion of local employment initiatives.

Widening the debate: the social dimension to future urban integration and cohesion

Previous debates on the need for a common European approach to urban policy have tended to focus on the transformation of the European urban system and how individual cities have coped with economic restructuring, globalization, the need for improved environmental awareness and the uncertainties caused by the introduction of the Single Market. However, in a period of continuing European integration a further question arises concerning the economic and social integration of disadvantaged groups particularly in the urban environment. The EU cannot hope to be successful in achieving economic prosperity, job opportunities and expansion of the Union without first securing internal cohesion. The growing risk for individual member states is that the economic, demographic and social pressures which are affecting the urban environment may accentuate the polarization of certain urban residents. As Padraig Flynn, the Commissioner for Employment and Social Affairs, stated in the introduction to the report entitled: 'Towards a Europe of Solidarity':

Europe faces many challenges, but one of the most difficult, which emerged in the 1970s and 1980s, is the changing nature and increasing incidence of poverty and deprivation. Whereas the Community entered the 1990s with a great deal of confidence, mainly on the foot of the prospects offered by the Single Market, at the same time, there were growing signs that the social situation of Europe's citizens might not be making equally positive progress. The effects of the world recession have intensified fears that social exclusion - a complex blend of interrelated factors - could become a persistent feature of the European social landscape (CEC, 1992b).

The economic and social processes that have shaped the urban environments of many European cities are likely to persist for the foreseeable future (European Foundation for the Improvement of Living and Working Conditions, 1986; 1989). Industrial decline and economic restructuring in the inner city and the outer housing estates has resulted in the combination of physical and social decline with high concentrations of unemployment in specific urban areas. As the EU develops a better understanding of urban change at the European level, it has also developed a better understanding of the impacts of change upon urban areas and urban communities. Research undertaken by the European Foundation for the Improvement of Living and Working Conditions (1986) suggested:

> The changing urban environment reflects the changing nature of national and international labour markets, changing fiscal priorities and changing life styles. For some groups this will enhance both living and working conditions. But for others, notably lower income, elderly and young single adults, single parent families and ethnic minorities, opportunities are likely to be constrained. A shrinking job market, the physical decay of the residential stock in many countries, public expenditure restraint and increased resources to market solutions, promise a more hostile urban environment for low income households...

The term which is most often used to describe these problems collectively is 'social exclusion'. The notion of social exclusion refers to the multiple and changing factors which result in people being excluded from the normal exchanges, practices and rights of modern society. Poverty is one of the most obvious factors which can indicate social exclusion, but social exclusion also refers to inadequate rights in housing, education, health and access to services. Social exclusion affects individuals as well as certain groups in society. It can be identified in urban and rural areas alike, as people are in some way subject to discrimination or segregation. Social exclusion, as used in the context of European policy, is a broad concept. It not only implies low material means but the inability to participate effectively in economic, social, political and cultural life. The Green Paper on Social Policy published in November 1993 was designed to encourage debate across Europe about the future direction and shape of European social policy (CEC, 1993b). It argued that Europe was experiencing a period of profound change. This was evident from the increase in globalization, the spread of new technology, changes in the organization of work, changes in Europe's population structure and the rise in costs of both health care and pensions. The Green paper concluded that unemployment in Europe was a structural problem and any foreseeable increase in Gross Domestic Product (GDP) in the community was not enough to tackle it. Any increase in productivity and wealth creation would need to be accompanied by measures to raise the overall level of employment in the European economy. The publication of the more recent White Paper on Social Policy - A way forward for the Union (CEC, 1994b) established the main strands of European social policy and insisted that competitiveness and social progress could flourish together. Although Europe requires a broad-based, innovative and forward-looking social policy, unemployment and the need for jobs is top of the policy agenda. Proposals on employment and training are an integral part of the policy response to unemployment initiated by the White Paper on Growth, Competitiveness and Employment (CEC, 1993c). The White paper has had important implications for the development of locally oriented polices designed to tackle unemployment and the problems caused by social exclusion.

Policies to promote economic and social cohesion in the urban environment

Europe's cities have been quick to realise the importance of funding from the EU. This does not mean that gaining financial support from the Commission is an easy option. Obtaining funds requires careful planning, good quality bids which reflect an understanding of current European priorities and regulations, and the need for matching funding from non EU sources (Chapman, 1995). The most significant source of European funding towards economic and social cohesion comes from the Community's Structural Funds. The EU's regional policies make explicit reference to urban communities as part of the criteria for Objective 2 eligibility. The Objective 2 designation covers regions in industrial decline and these are predominantly urban in character. Cities also receive support through Objective 1 programmes which include regions whose development is lagging behind the rest of the Community. With the latest round of European funding, the Structural Fund programming for 1994-1999, there is an increased emphasis on specific urban programmes under these two Objectives in a number of member states.

In addition, the urban dimension to EU action is increasingly mentioned, or directly targeted for financial support, by a number of EU programmes aimed at achieving sustainable development. For example, promoting integrated management models under the LIFE (Financial Instrument to Allow the Implementation of an Environmental Policy) programme; promoting urban energy networks and transport efficiency through the THERMIE (European Technologies for Energy Management) and SAVE (Special Action Programme for Vigorous Energy Efficiency) initiatives; improving telecommunications and advanced information technologies in cities under the RACE II (Research and Development Programme in Advanced Communications techologies in Europe), ESPIRIT (European Strategic Programme for Research and Development in Information Technology) and Transport Telematics programmes and finally assisting cities to promote economic and social cohesion through the use of European Regional Development Fund pilot schemes and the POVERTY III (Combat Social Exclusion) programme. Other European programmes with fewer financial resources offer opportunities for specific policy developments and include areas such as the Community Initiatives and the Human Resource Initiative which assists in employment training and opportunities for cities and urban areas to participate in European networks under the RECITE (Cooperation between regions and cities in Europe) programme. Of particular importance are the activities to develop cooperation between cities in the promotion of local Agenda 21 plans encouraged by the Sustainable Cities Project launched in 1993, as a follow up to the publication of the Green Paper on the Urban Environment in 1990.

Introduction of the new urban pilot programme (1995-1999)

On June 20th, 1995, the Commission established the new guidelines for the Structural Funds Innovatory Measures for the period between 1995-1999. This forms the second series of actions to be financed under Article 10 of European Regional Development Fund Regulations and includes support for Urban Pilot Projects (UPPs). The rationale behind this programme is the view that cities are today the main focus of economic growth and development, technological innovation and public service. At the same time, they all too often offer the worst examples of congestion, pollution, industrial decay and social exclusion. Cities in less-developed peripheral areas, as well as urban regions in decline, strive for economic development. Despite a steady reduction in disparities in per capita incomes between the regions, the gaps in development remain significant and illustrate the need to pursue with determination a policy of economic and social cohesion in the Community. In the period 1989-1993, the Commission, being aware of these problems, launched a series of 32 Urban Pilot Projects with the following four priorities:

• economic development in urban areas with social problems
• environmental actions linked to economic goals

19

- revitalization of historic centres
- exploitation of the technological assets of cities.

In 1994, following this experience, the Commission launched the URBAN (Balanced development of towns and cities) initiative, the aims of which were as follows:

to tackle the problems of high-risk neighbourhoods through an innovative and integrated approach resulting in actions which can be used as examples to be diffused in cities across the EU and to promote networks of exchange of experience and co-operation (CEC, 1994a).

For the period 1995-1999, the next series of UPPs is currently being launched. In the framework of a strategy for the development of urban areas, the general objective of the programme is to explore and implement innovatory avenues and thinking in urban policy and planning in order to contribute towards sustainable economic and social cohesion. Any urban, local or regional authority in the EU may participate in this programme provided it represents cities or conurbations with a population of over 100,000 inhabitants. Smaller towns could also be accepted, provided they have a marked urban economy and social structure, play a central role within a region, or are adjacent peri-urban conurbations of large cities. Measures under Article 10 could support pilot action across the EU, the features of which would be:

- the fact that they face problems common to a number of cities in similar conditions;
- the innovative and demonstrative character of the proposed solutions;
- the partnership between the public sector and other socio-economic partners is seen as an essential condition of their implementation, with a view to their becoming self-financing actions in the medium term.
- employment impact of the pilot actions (direct/indirect employment; safeguarding existing and/or creation of new jobs)

Proposals may relate to any activity of the participating local authorities provided that they are necessary components of a strategy agreed by the local partnership and aim to tackle specific urban problems, or anticipate future ones, in a sustainable way.

Themes, or combinations of themes, which come within the scope of the new UPP programme include:

- the improvement of urban planning of peripheral neighbourhoods in medium-to-large cities which have developed in an unplanned way

- the exploitation of cultural, geographical, historical or other advantages of medium-sized cities
- the regeneration of historic centres or deprived areas and the launching of new economic activities, or strengthening of existing ones, especially small and medium sized enterprises (SMEs), such as neighbourhood shops and craft enterprises, in combination with vocational training, rehabilitation, environmental actions or safety improvements
- tackling functional obsolescence in urban zones by introducing new uses which will provide needed urban or civic infrastructures, services or new economic activities
- improvement of facilities in downgraded urban districts around mass-transit stations
- the promotion of social/economic integration of minority groups and equal opportunity measures in particular through the establishment of partnership and citizen's participation
- steps to improve the environment through the creation of new open/green spaces and/or recreational activities together with sustainable facilities in built up districts
- integrated waste treatment and recycling activities
- reduction of energy consumption through renewable or clean alternative uses
- good practice for preserving building of architectural and social interest in regions with geographical disadvantages (eg earthquakes, floods)
- the integrated management of public transit/parking networks and enhancing the accessibility of isolated, economically disadvantaged neighbourhoods to labour markets
- the use of information technology in cities
- tackling institutional/legal issues necessary for the realization of innovatory schemes.

Conclusions

Since 1994 the Commission has introduced a more ambitious approach towards financing projects directly related to the urban environment. Given that the Commission has no competence in the field of urban affairs, these steps illustrate the growing emphasis upon the role of towns and cities in the process of European integration their importance in the achievement of economic and social cohesion. At the same time, the needs of spatial and urban planning have renewed the interests of policy makers in establishing a co-ordinated approach towards the physical, social and economic regeneration of the urban environment in a sustainable way. While the targeting of policy action at the urban level has been

21

considered to be subordinate to the wider objectives of the EU, the Commission has taken a more active role in recognising the specific problems experienced by urban residents. Actions undertaken by the EU in the urban environment have increased significantly since 1990, especially in the form of environmental policy, the Urban Pilot Projects, funding through the Structural Funds in Objective 1 and Objective 2 regions and through the introduction of the new URBAN Community initiative.

As concern for the future prosperity and spatial development of European urban areas continues, it is apparent that the EU will continue to develop and strengthen policies to improve the economic and social conditions in individual cities as well as developing networks of experience and interregional cooperation across the Community. The significance of the Community influencing urban policy must be examined in the wider context of the European policy agenda of the 1990s and the moves towards further economic, monetary and political integration. While there is growing evidence to support the case for urban areas to be a central part of EU policy, the Commission does not have a direct competence in urban policy or urban land use planning. The need to address complex economic, environmental and social problems which are present in many European cities may ultimately bring European policy into direct conflict with the member states and nationally based urban policy. Failure to reach a common perspective on the future spatial development of European cities could hinder the prospects for European integration and enlargement. Urban polices of national governments and the EU need to be more closely integrated, but instead of pursuing a suffocating equality, there should be room for diversity, and a recognition that what is good for the cities is good for the rest of the country and hence Europe (Dr Abraham Peper, Mayor of Rotterdam, City 2020, 1996, p.11).

References

Chapman, M. (1995), 'Urban Policy and Urban Evaluation: The Impact of the European Union', in Hambleton, R. and Thomas, H. *Urban Policy Evaluation: Challenge and Change*, Paul Chapman Publishing: London.

City 2020 (1996), *Reinventing the City*, City 2020: Leicester.

Commission of the European Communities (1990), *Green Paper on the Urban Environment*, CEC: Brussels.

Commission of the European Communities (1991), *Europe 2000 - Outlook for the Community's Territory*, CEC: Brussels.

Commission of the European Communities (1992a), *Urbanization and the Function of Cities in the European Community*, CEC: Brussels.

Commission of the European Communities (1992b), *Towards a Europe of solidarity. Intensifying the fight against social exclusion, fostering integration*, CEC: Brussels.

Commission of the European Communities (1993a), *Community Structural Funds 1994 -1999: Regulations and Commentary*, CEC: Luxembourg.

Commission of the European Communities (1993b), *Green Paper: European Social Policy Options for the Union*, CEC: Brussels.

Commission of the European Communities (1993c), *White Paper: Growth Competitiveness and Employment*, CEC: Brussels.

Commission of the European Communities (1994a), *Community Initiatives Concerning Urban Areas* (URBAN) COM(94) 61 Final\2, CEC: Brussels.

Commission of the European Communities (1994b), *European Social Policy. A way forward for the Union. A White Paper*. Com (94) Final, CEC: Brussels.

Commission of the European Communities (1994c), *Europe 2000+: Cooperation for European Territorial Development*, CEC: Brussels.

European Foundation for the Improvement of Living and Working Conditions. (1986), *Living Conditions in Urban Areas*, EFILWC: Dublin.

European Foundation for the Improvement of Living and Working Conditions (1989), *Social Change and Local Action: Coping with Disadvantaged in Urban Areas*, EFILWC: Dublin.

European Foundation for the Improvement of Living and Working Conditions (1994), *Bridging The Gulf*. EFILWC: Dublin.

Hall, P. (1994), 'Forces Shaping Urban Europe', *Urban Studies*, Vol. 30, No.2, pp. 883-98.

Mega, V. (1993), 'Innovations for the Improvement of the Urban Environment: A European Overview', *European Planning Studies*, Vol. 1, No.4, pp. 527-41.

Newman, P. and Thornley, A. (1996), *Urban Planning in Europe: International Competition, National Systems and Planning Projects*, Routledge: London.

Williams, R.H. (1996), *European Spatial Policy and Planning*. Paul Chapman Publishing: London.

23

3 Ecological modernization: a model for future urban and regional planning and development

Peter Roberts

Introduction

Even though the adoption of the principles of sustainable development in regional and urban planning has now become commonplace, considerable confusion and uncertainty remains regarding the implications of the adoption of this new code. In part, such anxieties stem from an absence of contextual knowledge regarding the origins, evolution and content of sustainable development concerns in planning. Prior experience can help to provide reassurance that a specific problem is not unique and may also yield possible solutions to current problems, but anxieties remain over the merits of alternative methods of implementing new commitments to what may often appear to be open-ended promises.

These uncertainties and confusions have led to the search for methods of translating the objectives of sustainable development into action in ways that avoid having to halt economic progress. Whilst there is considerable debate amongst analysts as to whether responsible economic progress can be considered to be an essential element of sustainable development, or whether its inclusion is simply an excuse for adopting a business as usual stance (Hajer, 1996), there is little doubt that many governments are unwilling to sacrifice the gains which they have obtained from previous eras of economic growth in order to pursue the goals of environmental enhancement. This implies that an important characteristic of a sustainable development policy, at least in the eyes of many governments, is that it should be capable of allowing economic progress to continue, albeit in a reduced or modified form.

This objective is not necessarily at odds with the goals of sustainable development, provided that a mode of economic progress can be identified and implemented which is in accord with the criteria defined for the attainment of sustainable - or balanced - development. It is this search that provides the central concern of this chapter. A further intention of the chapter is to cast light on the confusion which surrounds the terminology of sustainable development and, in doing so, to help to prepare an agenda for the guidance of planners as they make progress towards the adoption of a more sustainable model of development. This chapter commences with an investigation of the evolution of sustainable development concerns in regional and urban planning and, especially, the search for a balanced approach to the planning of regional and urban areas in which the requirements and importance of territorial integration are recognized. From this an attempt is made to define some of the major messages for the theory and current practice of planning, including the identification of appropriate objectives and goals for sustainable regional and urban planning and development. The third section of the chapter provides a more detailed examination of the merits of the concept of ecological modernization as an operational methodology for the implementation of planning strategies aimed at achieving sustainable development. The final section offers some initial conclusions and attempts to identify a number of key areas for future research and practice.

Sustainable regional and urban planning

Two converging strands in the development of theory and practice can be identified in the evolution of sustainability concerns in regional planning. The first of these strands - planning for sustainable development - can be seen as the most recent manifestation of the long standing search for a methodology of planning that will deliver balanced development. This represents a model of regional and urban planning in which the competing claims of the economy and the environment are assessed and judged by reference to previously identified and agreed criteria. A second line of evolution - from regional and development planning, through local economic development, to private and public investment - is associated with the recent emergence of the concept and practice of ecological modernization. Whilst these parallel strands of evolution represent the heritage of two different intellectual and operational traditions, they also draw upon shared concerns and display a number of common characteristics.

The origins of these common concerns and characteristics can be traced back to the contributions made by a number of the early planning and spatial theorists, and the work of the pioneers of regional and urban planning and development. Traditional analysts of localities and regions, such as Geddes (1915), considered that an appreciation of the importance of le Play's triad of folk, work and place, represented an essential step towards understanding the relationship between the environment, society and the economy. This relationship, which is fundamental to the concept of sustainable development, can also be seen as an expression of the mediating role of planning in seeking to achieve a balance between conflicting objectives and presenting an appropriate solution in a comprehensive, systematic and integrated manner.

In many discussions of planning for sustainable development relatively little attention has been paid to the social dimension. Despite the inclusion of specific social objectives and concerns in the Bruntland Report, and despite the recognition in that report of the consequences of a series of 'interlocking crises' (World Commission on Environment and Development, 1987, p.4), many observers and practitioners consider the challenge of sustainable development to be chiefly concerned with attempting to strike a balance between environmental and economic objectives. It is difficult to understand why questions of social equity and opportunity have been ignored, for the most part, in plans and strategies for sustainable development. This exclusion is all the more surprising given that a central objective of many such plans is the enhancement of the quality of life enjoyed by the inhabitants of an area.

The reason for highlighting the social dimension at this point in the chapter, is in order to make explicit the origins and consequences of the tensions which exist between the environmental, social and economic aspects of sustainable development. Returning to the fundamentals of sustainable development, the social and cultural politics of making choices regarding the future of the environment can be seen to represent the central dilemma facing many governments: is it possible to conceptualise and implement solutions to the environmental challenge of sustainable development within the existing social and economic system? In essence this represents a return to the previous debates on the characteristics and implications of the relationship between nature, technology and society (Hajer, 1996), and the outcomes of this debate have profound implications for planning.

Returning to the earlier discussion, it can be seen that one of the most important elements of the folk, work and place relationship is the value placed by individual members of society upon their personal and collective space or territory. This value is reflected in the strength of opposition that is increasingly mounted in defence of local and regional environmental assets - typically land in agrarian societies and environmental quality in urban areas. Whilst opposition to development on environmental grounds may be considered to be a luxury that can only be enjoyed by the inhabitants of economically successful western nations, there are signs that opposition to environmental degradation is also growing in less developed nations and regions. Growth at any price is no longer viewed as the only viable pathway to development, and accounting for the negative costs of the environmental damage caused by unthinking and excessive economic growth is frequently incorporated as a standard element in procedures for the approval of projects in both developed and less developed nations (Lutz and Munasinghe, 1991). Above and beyond the specific concerns which are expressed at regional or local level, increased care for the environment also reflects the need to protect the global commons (Blowers, 1993).

Economic growth which fails to respect the requirements of the environment also runs into the risk of economic sanction through trading restrictions, the withdrawal of borrowing rights and specific measures, including tariff-based actions that are aimed at preventing environmental dumping. Above all, ignoring the need to manage the capacity and quality of the environment can prove to be costly or even fatal (Cohen, 1993; Walker, 1994). Sustainable development can therefore be seen to be an essential component part of the planning and management of both developed and less developed nations and regions, especially under conditions which emphasise the considerable economic as well as environmental benefits available through the adoption of an ecological modernization model (Weale, 1993). This model argues that the future development of economic activities should proceed in-line with the needs of the environment. This topic will be developed further later in this chapter.

In the opinion of some planning researchers and practitioners, such concerns have always been at the heart of the planning agenda. Whilst the best planning practice - including a number of early attempts at the comprehensive development and management of both rural and urban areas in the USA during the 1930s, based upon the theory of balance development as expounded by the Regional Planning Association of America (MacKaye, 1928) - has always striven to achieve a balance between economic, social and environmental concerns, the majority of plans and schemes for regional and urban development have placed economic growth before balanced development. Such schemes have also tended to rely upon technology to resolve environmental problems when they occur. In many cases the adoption of a technocentric approach to planning and development was initially aimed at the elimination or reduction of resource constraints upon growth (O'Riordan and Turner, 1983), but few of these solutions have yielded lasting benefits and some treatments have actually proved to be worse than the problems that they were initially intended to cure.

The theory and practice of economic development has followed a similar pattern of evolution and has tended to place the achievement of successive increments of economic growth above the requirements of the environment. Many attempts to promote regional development through a traditional Fordist model have failed to stimulate either long term economic growth or the social and environmental well-being of an area (Stohr, 1989), and such models have also frequently regarded the resources of a region as a storehouse to be plundered at will. Perloff and Wingo (1964) provided an early assessment of the relationship between economic growth and the environment in which they argued that because the resource endowment is continually redefined by changes in final and intermediate demand, production technology, and economic organization, factors other than those associated with the best environmental solution exert considerable influence upon the pattern of

27

resource development. In this traditional model of development, resources and the environment were utilized as the basis for local and regional specialization. A similar paradigm of economic development has dominated much of western thinking with regard to advice and assistance to less developed nations, thereby perpetuating what Friedmann and Weaver (1979, p.78) condemn as a process of 'resource development as a means of economic expansion'. This application of the traditional model has emphasized the relatively narrow requirements of economic growth rather than the balanced social, economic and environmental uplifting of a people that is the distinctive characteristic of development.

An additional feature of the traditional growth-dominated model that is common to both planning and economic development, is the domination of functional integration over territorial integration. Whilst it might be expected that regional and urban planning and development would give priority to the provision of spatial coherence, in practice many such exercises have failed to provide the specific guidance necessary in order to direct public and private policy and investment towards the achievement of a solution which is based upon the enhancement of territorial integration. Many regions have been planned and developed in a manner which is principally designed to serve the needs of national growth, sectoral requirements, or the desires of multinational companies to maximise profit. Although this model of regional development as a form of either internal or external colonialism is well-known, the implications for the environment have generally been ignored. If regions are to function as social, economic and ecological entities, it is essential to encourage and ensure the maximum degree of territorial integration and coherence. This argues in favour of integrated and comprehensive territorial or spatial planning, the definition of administrative regions that are as far as possible co-terminus with natural regions, and the adoption of planning and development policies that place particular emphasis on the need for balanced development. These conditions are similar in many respects to the principles advocated for the overall guidance of the future spatial evolution of the European Union (Commission of the European Communities, 1994) which, when integrated with the principles agreed for sustainable development (Commission of the European Communities, 1992), can help to define an operational agenda for regional and urban planning.

Planning approaches and objectives

Although it is acknowledged that a number of previous attempts at regional and urban planning and development have attempted to encapsulate the concerns inherent in the concept of sustainable development, it is also apparent that much of the past and current practice has been chiefly concerned with the achievement of other objectives. This is true in both east and west. In the case of the less developed nations Drakakis-Smith has referred to the spurious nature of the relationship between urbanization and development, often based upon the 'old and false surrogate GNP per capita' (Drakakis-Smith, 1995, p.660), whilst in the west Simonis (1993) has identified the continuing obsession with 'tonnage ideology' and has argued that a more fundamental reappraisal of the socio-economic objectives of advanced nations is required before sustainable development can be tackled with the necessary degree of determination.

This suggests that it is essential to clarify and re-state the concept of sustainable development as it applies to regional and urban planning, and to identify operational procedures and criteria that may be used to guide its achievement. The most obvious entry point into this debate is through the work of the World Commission on Environment and Development (1987) and its definition of sustainable development as 'development that meets the need of the present without compromising the ability of future generations to meet their own needs' (ibid., p.8). Jacobs (1991) and Healey and Shaw (1993) share a similar approach to the definition of the three essential components of this concept:

- the inclusion of environmental concerns in economic policy making;
- intergenerational and intragenerational equity;
- the need to adopt a balanced model of development rather than an approach that is aimed solely at the promotion of economic growth.

A working definition of sustainable development should aim to distinguish between sustained growth and sustainable development, should incorporate social welfare as well as environmental concerns, and should attempt to provide the basis for the achievement of a balanced development solution. In addition, it is important to avoid falling into the trap of adopting a working definition that is so rigid that it denies the reality of 'pragmatic intrusions' (Drakakis-Smith, 1995, p.662), or which ignores the considerable diversity of regional situations that require the construction of specific models in order to achieve effective sustainable development. This spatially-specific mode of construction is especially important in order to allow operational change in the ecological sphere to be institutionalized in the 'social practices and institutions of production and consumption' (Mol, 1995, p.30). Although, as will be discussed in the third section of this chapter, such ideas have their origins in the restructuring, or ecological modernization, of the manufacturing sector, they are especially valid in a discussion of the production of the built environment.

Extending these ideas further, it can be seen that the defining characteristics of sustainable development argue in favour of the adoption of a model of planning and development that is based equally upon economic, social and environmental objectives, rather than the limited vision observable in many economic interpretations (Serageldin, 1993). Ecological modernization shares many of the

29

characteristics of sustainable development and places particular emphasis on the design and operation of a political programme favouring a set of policies aimed at ensuring that environmental protection is not a burden on the economy (Weale, 1993). It can be seen as an operational mechanism for the achievement of sustainable development by suggesting that it is possible to integrate the goals of economic development and environmental protection. Through this reconciliation, it is argued, synergies will be generated which can be harnessed and put to good use (Gouldson, 1995).

Equally important in the design and implementation of planning and development for sustainable development is the vital role played by strategy. Thinking about long-term needs and the likely evolution of economic, social and environmental conditions, is a prerequisite for effective planning, and this is even more so the case if the intended output from the planning process is a plan which matches the requirements of sustainable development. For most of the past two centuries the proponents of economic growth, especially in the western world, have regarded the environment as a storehouse of resources and as a free dumping ground for waste. Changing course will take time, and even though it is vital to take immediate action in some cases in order to prevent further environmental degradation, strategic planing for sustainable development is unlikely to be fully effective in less than a generation. This implies a minimum time horizon of twenty five years, way beyond the normal target date for many public policy activities (Begg, 1991). Indeed it can be argued that short-term palliatives are inherently dangerous because they generate complacency and the belief that an environmental problem can be solved once and for all.

In designing a model of planning and development that encapsulates sustainability characteristics and objectives, it is important to recognise the need to design solutions in order to fit the particular economic, social and environmental conditions evident in an individual region or locality (Roberts, 1995). Places differ, and even though it is possible to transfer the principles of sustainable planning from one region to another it is vital that the particular inheritance of economic and environmental factors evident at an individual location should be acknowledged as the most appropriate foundation for policy. Bottom-up, as against top-down, solutions are a distinguishing characteristic of most successful attempts at planning for sustainable development (Wood, 1994). This lesson reflects the previous experience of development from below (Stohr and Taylor, 1981) and is advocated in order to ensure a high degree of social ownership of any suggested planning solutions.

A final element of considerable importance in the development of an approach to sustainable regional and urban planning, is the need to ensure the integration of policy and the desirability of embedding environmental concerns across the full span of sectoral activities. This is a difficult task in a single region or locality, and is even more challenging at a national or transnational level where, for example, responsibility for a particular function is likely to be vested in different sorts of agencies with varying degrees of power and responsibility (Sbragia, 1992). A wider and broader manifestation of this issue is to be found in situations in which cultural or ideological differences have created political or conceptual barriers to the establishment of mutually beneficial solutions to the occurrence of environmental problems. This was certainly the case in the former Soviet Union - during the Stalinist period, 'nature was considered as an obstacle to further progress' (Tellegen, 1996, p.76) - and similar instances, although on a lesser scale, continue to obstruct the implementation of common solutions to transboundary problems.

Having reviewed some of the more important issues involved in establishing a sustainable approach to planning and development, it is possible to identify a number of characteristics that might be expected to figure as major elements in such a model. In the view of Drakakis-Smith (1995) these include the following requirements:

• equity, social justice and human rights;
• basic human needs;
• social and ethnic self-determination;
• environmental awareness and integrity;
• awareness of inter linkages across both space and time.

Because of the role assigned to it in most advanced societies, but heeding the warning sounded by Szerzynski, Lash and Wynne (1996) regarding the dangers of attempting to continue to accommodate economic growth objectives in sustainable development strategies, regional and urban planning has also to:

• identify ways of adapting and adjusting the economic base of society in order to conform to the new requirements of sustainable development; and
• ensure that future development avoids the pitfalls of the past.

At a more focused level of attention, and in relation to the environmental component of sustainability, the specific requirements of a conventional planning policy and operational system would include, amongst others, those factors specified by the OECD (1990) and Roberts (1994). Particular emphasis should be placed upon:

• developing and monitoring long-term strategies in order to satisfy the broader goals encapsulated within the concept of sustainable development;
• adopting a cross-sectoral approach in order to integrate policy responses and to embed environmental concerns in other fields of development activity;

31

- facilitating co-operation between all major public, private and voluntary sector organizations and agencies;
- setting and enforcing minimum standards of environmental performance in all sectors;
- enabling producers and consumers to capture and absorb the negative environmental costs associated with the production and use of goods and services;
- encouraging local and regional initiatives;
- minimising costly and environmentally inefficient transfers of resources and goods;
- developing and applying models and methods of land-use and other resource allocation activities that can help to enhance accessibility and minimise unnecessary travel;
- generating hard and soft infrastructures that enable the preceding objectives to be achieved.

Finally, and above all, it is important to screen and review development proposals in order to ensure that new activities do not add to the existing stock of environmental problems or generate negative externalities. This is a matter of concern at all levels in the policy and operational hierarchy. The case for introducing strategic environmental assessment (SEA) has been argued for some considerable time (Glasson, 1995; Therivel et al., 1992), and the adoption of SEA is now acknowledged as a vital step in avoiding further environmental degradation and in reducing the waste of economic and human resources associated with the generation of projects that fail at the final hurdle to satisfy important sustainability criteria.

Ecological modernization strategies for planning

The previous sections of this chapter have reviewed the origins, development and content of attempts to provide a balanced approach to regional and urban planing, have considered the failure of planning to achieve these ideals, and have attempted to identify the changes necessary in order for planning to operate in a manner which satisfies the requirements of sustainable development. In the final section of the chapter attention is focused on the validity of the ecological modernization model (or models) as a foundation for the establishment of a more sustainable mode of planning

Much of what constitutes an ecological modernization approach has been previously tried and tested within planning. The pioneers, such as Geddes and MacKaye, sought to work with the grain of the environment in the advancement of economic and social progress. However, the best practice of the past has frequently been dominated by the mediocre, often with a result that has failed to satisfy the economic, social or environmental objectives originally specified and agreed.

A parallel failure in the manufacturing sector to come to terms with the requirements of the environment, and the realization of the consequences of such failure, has led to the emergence of the concept and practice of ecological modernization. As noted in the previous section of this chapter, ecological modernization suggests that it is possible to integrate the goals of economic development with those of environmental protection. However, whilst this concept appears at first to offer a pain free routeway to the achievement of sustainable development, in reality choices and compromises still occur and require resolution.

Hajer (1996) identified three different interpretations of ecological modernization:

- as institutional learning - in which industrial and administrative organizations learn from the critiques of conventional industrial society put forward by the environmental movement, and from this develop modes of economic development which are compatible with the environment - this interpretation assumes that existing institutions can internalise ecological concerns;

- as a technocratic project - in which radical environmentalists argue that the ecological crisis requires more than social learning by existing organizations and that ecological modernization as institutional learning offers a false solution to very real problems - this interpretation argues that without fundamental structural changes in the economy, environmental problems are left unaddressed;

- as cultural politics - this reading of ecological modernization sees the debate on environmental problems as a reflection of wider debates on the preferred social order, this interpretation emphasises the nature and processes of social and political debate about what sort of society - and environment - we want and how to negotiate social choice in order to arrive at the preferred option.

From these three interpretations it is possible to identify three distinct and distinctive modes of response. Each of these responses represents a stage towards what Simonis (1989, p.358) referred to as 'the necessary and feasible harmony between man and nature, society and environment.'

The critique advanced in Hajer's second interpretation - ecological modernization as a technocratic project - points to the inherent weakness of end-of-pipe solutions, or to the assumption that new technologies can be relied upon to counter new environmental damage. This implies that positive action, based on the precautionary principle and the desirability of ensuring that prevention rather than cure is the guiding principle of practice, should be emphasized. Whilst the merits of this second interpretation are self-evident, they do not, in themselves, provide a basis for progressive change in most western societies. Rather, it is the first and third of Hajer's interpretations that represent the most fertile soil in which to cultivate a new mode of regional and urban planning.

The first interpretation - ecological modernization as institutional learning - is the most commonly adopted mode of reorientation. At the core of this interpretation is the need to break with the past in order to move away from a reactive mode of response to environmental problems and towards a mode of operation that avoids the need for reaction through the adoption of a greater sense of anticipation. This implies, in Hajer's (1996) view, that environmental degradation should not simply be viewed as an external problem, instead environmental concerns should be integrated within policy-making. The use of procedures such as SEA can help to ensure that such integration occurs.

However, despite the substantial progress which has been made in the manufacturing sector in some countries towards the achievement of ecological modernization that is in accord with the conditions of the first interpretation, further progress inevitably means moving towards the conditions specified by Hajer in his third interpretation - ecological modernization as cultural politics. The third mode suggests that there are choices to be made as to 'what sort of nature and society we want' (Hajer, 1996, p.259). In defining and determining these choices the role of discourse is central. Through discourse future scenarios can be debated and constructed and, having demonstrated the implications of such scenarios, alternative goals and pathways can be identified. This mode of interpretation and operation is similar in some respects to that embodied in a strategic vision mode of analysis and planning (Roberts, 1990).

What results can be identified from the practice of ecological modernization and what messages can be gleaned that may be of help in determining the future of regional and urban planning? There are a number of specific case studies of ecological modernization projects. Mol's research into the adoption of ecological modernization in the chemical industry demonstrates 'how the environment moves into the process of chemical production and consumption, and transforms it' (Mol, 1995, p.391). This transformation is a slow process, more in accord with Hajer's first interpretation, but nevertheless demonstrates the value of the integration of environmental objectives into corporate goals and production strategies, and the growing coherence of responses. A second case study is of the wider restructuring project undertaken in the former German Democratic Republic, a project that has placed ecological modernization at the centre of a series of economic, social and environmental actions aimed at responding to the degraded conditions brought about by the 'erroneous ways of the unchained, large scale

34

technology of chemical and energy production' (Kiegler, 1994, p.22). The goals of the project implemented in the Dessau - Bitterfield - Wittenberg region are to:

• review the architectural and cultural heritage;
• regenerate natural resources;
• reject technological cul-de-sacs, polluting production and 'tonnage' ideology;
• move away from production for distant markets and orientate production, through research and innovation, towards regional needs;
• develop regional economic networks; and
• enhance training with the needs of new forms of activity in mind.

The overall intention is the long term restructuring of the region in order to ensure its economic, social and environmental future.

Reflecting the results of these and other projects, the implications for regional and urban planning can be seen to be related to three issues. First, the need for planning to be considered as the outcome of a new form of discourse on the broader goals of society. Second, the desirability of ensuring the fullest possible integration between the goals of sustainable development and the general goals of planning. Third, the need for practice to be based on realistic assumption as to what can be achieved. Each of these issues is considered in more detail in the following section of this chapter.

Towards a new model for planning

This chapter has investigated the origins, key concerns and practice of regional and urban planning in relation to sustainable development. In addition, it has considered the potential role of ecological modernization as an organising concept in the further development of planning theory and practice.

From the evidence presented, it is apparent that much of the previous theory and practice of regional and urban planning has failed to deliver sustainable (or balanced) development. An explanation for this failure is provided in Hajer's second interpretation - planning has frequently resorted to technocratic solutions to environmental problems and has failed to address the structural aspects of the problem. This failure is not simply a failure of planning, rather it is a failure of the role defined for planning by society at large.

This returns the discussion to the three issues presented at the end of the preceding section. Regional and urban planning is a reflection of the broader socio-cultural values which it seeks to deliver. If these values ignore or diminish the challenge of sustainable development, then planning itself is powerless to bring about positive change. In such a situation the best that can be expected is that planning will seek to moderate the most extreme outcomes of social choices that work against the attainment of sustainable development.

However, set against this dismal prospectus is evidence which suggests that regional and urban planning is beginning to move towards the integration of the goals of sustainable development within the mainstream of theory and practice. This suggests that institutional learning has taken place and that further progress will be made as evidence accumulates which demonstrates the 'presence of a synergistic relationship between economic development and environmental protection' (Gouldson, 1995, p.6). If this is the case, and the current evidence suggests that it may be so, then Healey and Shaw are correct in their assumption that the 'realization of environmentally sustainable strategies is not simply a problem of technology or ecosystematic understanding, but of politics and the articulation and implementation of public policy' (Healey and Shaw, 1993, p.772).

As a consequence, and mindful of the long lead times involved in the realignment of cultural politics, the role of regional and urban planning can be defined as a proactive mechanism that has the task of helping to provide leadership in the translation of sustainable development theory into practice. The key items for practice specified at the end of the second section of this chapter reflect the conventional challenge, whilst a more adventurous agenda would reflect the tasks specified by Busch-Luty (1995, p.24) which include:

- establishing regional agendas for sustainable development which put detailed flesh onto the bones of national/international agreements and which co-ordinate and integrate bottom-up approaches;
- introducing a new socio-cultural paradigm in order to allow for democratic choice;
- taking into account the complex interactions between the economic, social, ecological and cultural dimensions of the new paradigm, and applying the resulting 'collective consciousness' in a homogenous ecological region;
- diminishing the power of the 'industrial and authoritarian state' as the impetus for growth;
- returning economy and society to 'naturally necessary limits' by identifying ecologically correct process;
- conserving resources through strategies for durability, re-use and recycling; and
- emphasising independent human development.

This chapter has sought to offer some reflections on the debate on sustainable development as it relates to the theory and practice of regional and urban planing. It has avoided offering trite solutions in order to enable the need for greater understanding to be emphasized. Although substantial progress has been made in recent years in accommodating new concerns, and this includes a return to the search for a sustainable (or balanced) mode of planning practice, there is still a considerable task to be undertaken. Perhaps the next step will have to be delayed until society makes its intentions clear, but is it possible to wait that long?

References

Begg, H. (1991), 'The Challenge of Sustainable Development', *The Planner*, 77, 22, pp. 7-8.

Blowers, A. (1993), 'Pollution and Waste - a Sustainable Burden?', in Blowers. A. (ed.), *Planning for a Sustainable Environment*, Earthscan: London.

Busch-Luty, C. (1995), 'Sustainable Development in the Tension Between the Global Economy and the Regions', *Bauhaus- Forum* 1995, Bauhaus: Dessau.

Cohen, M. (1993), 'Megacities and the Environment', *Finance and Development*, 30, 2, pp. 44-7.

Commission of the European Communities (1992), *Towards Sustainability*, Commission of the European Communities: Brussels.

Commission of the European Communities (1994), *Europe 2000+*, Commission of the European Communities: Brussels.

Drakakis-Smith, D. (1995), 'Third World Cities: Sustainable Urban Development', *Urban Studies*, 32, 4-5, pp. 659-77.

Friedmann, J. and Weaver, C. (1979), *Territory and Function*, Edward Arnold: London.

Geddes, P. (1915), *Cities in Evolution*, Williams and Norgate: London.

Glasson, J. (1995), 'Regional Planning and the Environment: Time for a SEA Change', *Urban Studies*, 32, 4-5, pp. 713-31.

Gouldson, A. (1995), *Ecological Modernization and the European Union*, Paper Presented at the European Environment Conference: Nottingham.

Hajer, M. A. (1996), Ecological Modernization as Cultural Politics, in Lash, S., Szerszynski, B. and Wynne, B. (eds), *Risk, Environment and Modernity*, Sage: London.

Healey, P. and Shaw, T. (1993), 'Planners, Plans and Sustainable Development', *Regional Studies*, 27, 8, pp. 769-76.

Jacobs, M. (1991), *The Green Economy*, Pluto Press: London.

Kiegler, H. (1994), 'Bitterfield - Without Memory', *Bauhaus-Forum* 1994, Bauhaus: Dessau.

Lutz, E. and Munasinghe, M. (1991), 'Accounting for the Environment', *Finance and Development*, 28, 1, pp. 19-21.

MacKaye, B. (1928), *The New Exploration: A Philosophy of Regional Planning*, Harcourt Bruce: New York.

Mol, A.P.J. (1995), *The Refinement of Production*, Van Arkel: Utrecht.

Organization for Economic Co-operation and Development (1990), *Environmental Policies for Cities in the 1990s*, OECD: Paris.

O'Riordan, T. and Turner, R.K. (eds) (1983), *An Annotated Reader in Environmental Planning and Management*, Pergammon: Oxford.

Perloff, H. and Wingo, L. (1964), 'Regional Resource Endowment and Regional Economic Growth', in Friedmann, J. and Alonso, W. (eds), *Regional Development and Planning*, MIT Press, Cambridge: Mass.

Roberts, P. (1990), *Strategic Vision and the Management of the UK Land Resource*, Strategic Planning Society: London.

Roberts, P. (1994), 'Sustainable Regional Planning', *Regional Studies*, 28, 8, pp. 781-7.

Roberts, P. (1995), *Environmentally Sustainable Business: A Local and Regional Perspective*, Paul Chapman: London.

Sbragia, A. (1992), *Environment Policy in the European Community: The Problem of Implementation in Comparative Perspective*, Round Table Seminar Towards a Transatlantic Environmental Policy, The European Institute: Washington.

Serageldin, I. (1993), 'Making Development Sustainable', *Finance and Development*, 30, 4, pp.6-7.

Simonis, U.E (1989), 'Ecological Modernization of Industrial Society: Three Strategic Elements', *International Social Science Journal*, 121, pp. 347-61.

Simonis, U. E. (1993), *Industrial Restructuring: Does it Have to be Jobs Vs. Trees?* Work in Progress of the United Nations University, 14, 2, p.6.

Stohr, W.B. (1989), 'Regional Policy at the Cross-roads: An Overview', in Albrechts, L., Moulaert, F., Roberts, P. and Swyngedouw, E. (eds), *Regional Policy at the Cross-roads: European Perspectives*, Jessica Kingsley: London.

Stohr, W.B. (1989) and Taylor, D.R. (1981), *Development from Above or Below?*, Wiley: Chichester.

Szerszynski, B. Lash, S. and Wynne, B. (1996), 'Introduction: Ecology, Realism and the Social Sciences', in Lash, S., Szerszynski, B. and Wynne, B. (eds), *Risk, Environment and Modernity*, Sage: London.

Tellegen, E. (1996), 'Environmental Conflicts in Transforming Economies: Central and Eastern Europe', in Sloep, P. and Blowers, A. (eds), *Environmental Problems as Conflicts of Interest*, Arnold: London.

Therivel, R., Wilson, E., Thompson, S., Heaney, D. and Pritchard, D. (1992), *Strategic Environmental Assessment*, Earthscan: London.

Walker, G. (1994), 'Industrial Disasters, Vulnerability and Planning in Third World Cities', in Main, H. and Williams, S.W. (eds), *Environment and Planning in Third World Cities*, Wiley: London.

Weale, A. (1993), 'Ecological Modernization and the Integration of European Environmental Policy', in Liefferink, J.D., Lowe, P. D. and Mol, A.P.J. (eds), *European Integration and Environmental Policy*, Belhaven: London.

Wood, C. (1994), *Painting by Numbers*, Royal Society for Nature Conservation: Lincoln.

World Commission on Environment and Development (1987), *Our Common Future*, Oxford University Press: Oxford.

4 Spatial and environmental problems of border regions in East-central Europe, with special reference to the Carpathian Basin

József Tóth and Pál Golobics

Introduction

In our opinion, regional investigations differ in certain respects from traditional geographical descriptions of areas. Activities involved in the consideration of the various components need to take account of the inter-relations between social, economic, infra structural and natural spheres. There comes a time when the course of the development of the productive forces reaches a stage in which regions become internally distinct and segregated from each other according to aspects of the division of labour and other characteristics. These regions have individual developments, structures and problems which give them an inner cohesion and identity. The texture of these characteristics in the core region is dense and it becomes diluted towards the periphery. Eastern-Mid-Europe started to reach this stage in the mid-19th century.

The interrelations of regions in this situation are based on the fact that because of economic specialization an exchange of activities and goods develops, not only within, but more particularly between regions. Although relationships between certain regions come into being at an early stage of the development of the productive forces of society, definitive interregional relationships can be identified only at the more developed stage of regional formation. This could be considered as a reality in Eastern-Mid-Europe by the end of the 19th century. We argue that regions are parts of the country influenced and created by the secular development of social and economic patterns, and by the configuration and differentiation of structure. Some of the borders of the regions considered here partially coincide with the borders of the country at the same time. However the geographical reality of the regions is not necessarily changed if they are cut across by borders temporarily reflecting international power relations. In this sense the areas become international, or trans-boundary regions. At the same time it is natural to conclude that the functioning of these regions is fundamentally influenced by the social and economic relations of the adjoining states and the political willingness for cooperating with each other. In the case of the international regions, the intra- and inter-regional cooperation thus becomes a deeply political question. In our view there is increasing support for intra- and inter-regional cooperation across national borders in Eastern-Mid-Europe today. This trend is not yet fully established in strategies of the nations involved, but it is being encouraged by the willingness of these states to align themselves with, or even to become members of the European Union.

Historical references

The situation before 1914

There are many events in the history of the Carpathian basin that variously divided, combined and restructured the population and the land for longer or shorter periods. Interestingly none of these were built on natural geographical differences. If these periods are even glanced through, it will become obvious that the changes were largely the results of military-political reasons and that agreements and alliances were exceptional, occasional and partial. By no means can they be considered as harbingers of regional developments, or to represent the formation of regions.

Amongst early examples of the territorial units that were temporarily formed in the Carpathian basin are:- provinces of Roman Age (Pannonia, Dacia); territories belonging to the time of the Hungarian settlement, areas of medieval defence-systems surrounding the country with a ring of uninhabited and scarcely penetrable marches and waste land; different divisions of Hungary and Croatia and then Hungary and Transylvania for the 150-year split of Hungary followed by the Turkish invasion. These temporarily fixed borders did not coincide either with each other or with natural features, or ethnic borders, so they were neither long lasting enough nor characteristic enough to become significant frameworks for regional development. Similar statements could be made in connection with Joseph II's short-lived territorial districts, or with the territories created for early enumerations. The units of the thousand years old country system are too small and their borders are too insignificant for them to be taken into account as regional units. Military, administrative and juridical units possessed only one function and were thus too narrowly based to function as coherent regions. Borders created for different reasons invariably ignored the living-space of ethnic groups, and criss-crossed them.

Although there were no secularized developments partitioning the Carpathian basin into regional units, the unity of the Carpathian basin area itself is clear. Its principle expression, based largely on natural features, is the historical state of Hungary, reorganized by the settlement of 1867 within the Dual Monarchy. In spite of the fact that from 1868 Croatia possessed political autonomy and Transylvania had its own conscious independence, Hungary can be considered an entity, as depicted in Figure 4.1.

Figure 4.1 Borders during the Austro-Hungarian monarchy

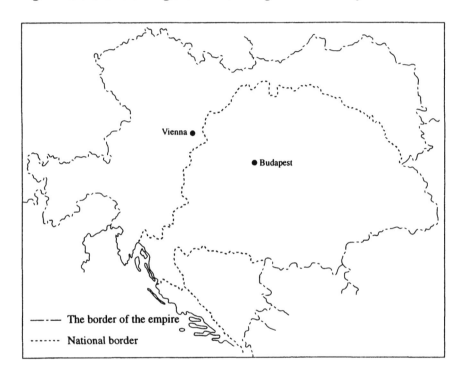

In the nineteenth century, this unity was consolidated by further developments, notably the unified railway network (Figure 4.2). Its density decreases everywhere towards the borders except where it shows a transition, indicating a unified development, towards Austria.

By the beginning of the 20th century in historical Hungary certain distinctive regional cores were emerging and, through the process of economic development, definitive regions became established. In between them, less densely textured zones extended, with multilateral and less intensive relationships. Over time, these transitional zones become integrated with the regional framework. In our view there were nine initial regions in Hungary after the turn of the century (Figure 4.3).

Figure 4.2   Railway network of Hungary in 1913

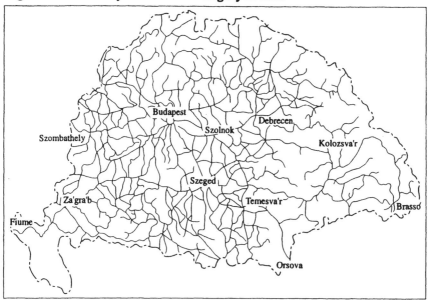

Source: The History of Hungary, Vol.4.

Figure 4.3   Schematic view of regions in the Carpathian Basin in the early 20th century

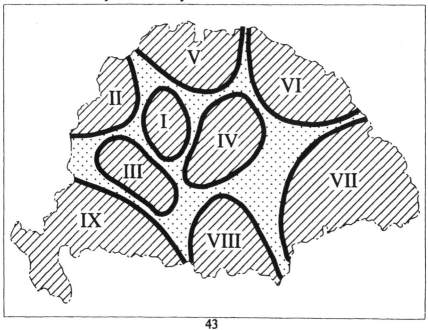

43

Among these, the premier political and geographical core area was constituted around Budapest because it was a focal point of natural endowments, and had an excellent transport-geography situation. A similar development characterized the second area that was organized around the common capital, Vienna, namely Small Plain and Western Transdanubian, because this metropolis is peripherally situated in Austria. The third area, Middle and southern Transdanubian, had a looser texture and was organized around smaller towns. The Great Plain, the fourth area, was characterized by a lower level of regional development, being an agricultural area with some significant cities, a transport network with great capacity, but altogether with an underdeveloped infrastructure.

Upper Hungary was the fifth area, and the sixth one was Ruthenia. These were less clearly integrated regions that stretched up over the mountain areas and contained a predominance of non-Hungarian ethnic groups.

Transylvania is the seventh area to be considered. Its evolution into a region was based on history and was supported by a coherent identity. Its economic system and mixed Hungarian, Romanian and German population made it an interesting region. Although it had close relations with the centre of the country, it simultaneously had significant Moldavian links.

The eighth area constituted the South, sometimes called the 'groin' of Hungary. It was an area with excellent agricultural conditions, mixed ethnicity (Hungarian, Serbian, German and Romanian) an unambiguous central orientation, but open towards the South. It also had significant Balkan relationships, and a noticeable attraction towards Belgrade.

The ninth area, Croatia possessed the political framework of independence. Its historical past, complex ethnicity, maritime and other relations make it undoubtedly the most independent region, in spite of the attachments with Budapest.

In summary, the pre-1914 Hungary, that filled the Carpathian Basin could be divided into tentative regions with different levels of developments but these could not be interpreted as definitive regions.

The interwar period

Political re-arrangements following the first world war brought many new borders into the Carpathian Basin. These political borders (Figure 4.4), followed neither ethnic nor natural divisions but were exclusively the results of power politics. They took little account of the emerging regions outlined above. The new borders cut the regions into two (sometimes into more) parts, and in many cases inter-regional relationships became impossible.

Regional developments also met difficulties when the whole of Middle Europe, including the Carpathian Basin, was broken up into little parts. New states came into existence, and the general atmosphere of hostilities was not favourable to international regional cooperation across the borders. These many borders (Figure 4.5) slowed transport, made the 'working' of the relationships more expensive, and necessitated many new patterns of movement.

Figure 4.4   Hungary after the First World War

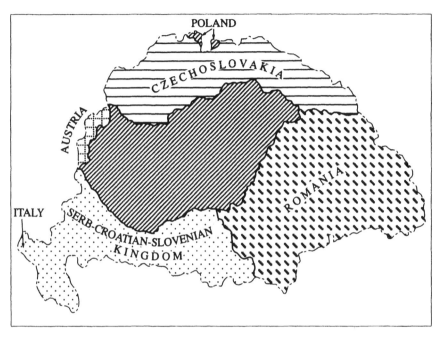

Source: Census of 1920. VI. Part Budapest, 1929.

Political borders crossing the Carpathian Basin made what had previously been internal regional relationships into international links in a legal sense. This circumstance was not necessarily a social or economic impediment, because in many cases states were in close contact, living at peace with each other and with good transport links, so regional relationships could continue as they did in regions not disturbed by political borders. However, in other cases state borders in the Carpathian Basin truncated social and economic linkages on both sides of the border and acted as a serious impediment to development. This regrettable situation was only modified where border stations were operating and regional relations and also development potentials were concentrated and enhanced (Figure 4.6).

So political realignments and oppositions between the two world wars acted as significant impediments in the processes of regional development and regional formation in the internationalized area of the Carpathian Basin.

Figure 4.5 Geo-political changes in Central Europe

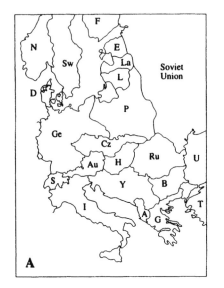

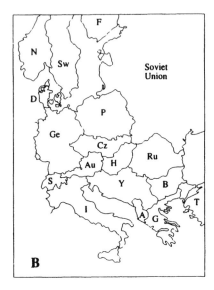

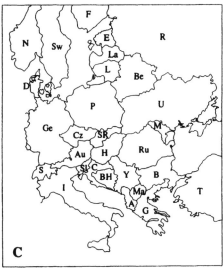

Geo-political changes in Central Europe

A Situation between the two war

B Situation after World War II

C Current situation (1997)

A	Albania	L	Lithuania
Au	Austria	Ma	Macedonia
Be	Belorussia	M	Moldavia
BH	Bosnia-Herzogovinia	N	Norway
B	Bulgaria	P	Poland
C	Croatia	Ru	Rumania
Cz	Czech Republic	R	Russia
D	Denmark	SR	Slovak Republic
E	Estonia	Sl	Slovenia
F	Finland	Sw	Sweden
Ge	Germany	S	Switzerland
G	Greece	T	Turkey
H	Hungary	U	Ukraine
I	Italy	Y	Yugoslavia
La	Latvia		

Figure 4.6   Forms of regional connections

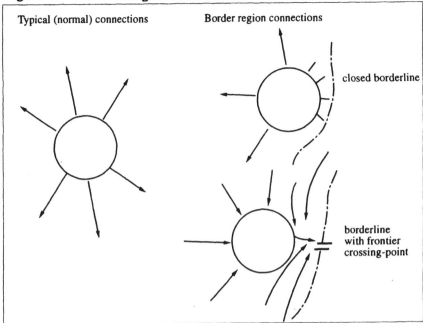

Typical (normal) connections

Border region connections

closed borderline

borderline
with frontier
crossing-point

The period of state-socialism

Peace concluding the second world war broadly re-established the pre-war situation, except for a widening of the abutment of Bratislava and the annexing of Ruthenia to the Soviet Union. The appearance of the Soviet Union in the Carpathian basin and the political and economic power that it exerted over the whole of Eastern-Mid-Europe also had a widespread influence upon the regional development of the Carpathian Basin.

Southern-Mid-European states, led by the Soviet Union, provided friendly and even fraternal relations at the level of official declarations. However, by over-emphasizing their territorial integrity and the principle of non-intervention into internal affairs, by glossing over ethnic problems, and through the permanent operation of highly centralized political and economic structures, they created a situation in which borders were handled as fetish, and they became difficult to cross. Cross-border co-operation between areas that had developed uniformly for centuries was possible with the knowledge and permission of the capitals, but it was frequently difficult and instead of the normal single step there were five actions necessary for it to take place (Figure 4.7).

Smaller countries often had stronger relations with the Soviet Union than with each other. This fact created an increasing isolation along the borders, and made inter-regional co-operation very difficult.

Figure 4.7   Communication pathways across a closed national border

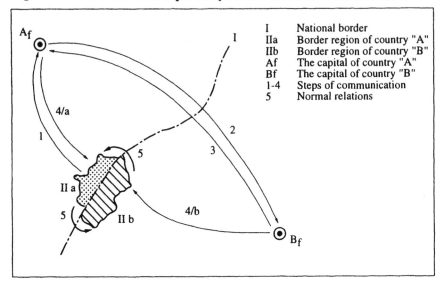

I	National border
IIa	Border region of country "A"
IIb	Border region of country "B"
Af	The capital of country "A"
Bf	The capital of country "B"
1-4	Steps of communication
5	Normal relations

Figure 4.8   Configuration of the Hungarian settlement system

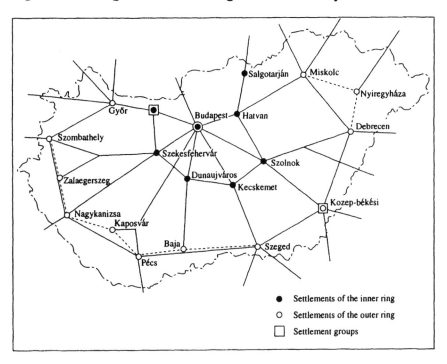

- ● Settlements of the inner ring
- ○ Settlements of the outer ring
- □ Settlement groups

The regional structure of the states behind the new political borders within the Carpathian Basin developed within strict political limits. One consequence was that the political borders became stronger and more meaningful than the traditional regional borders, since politics in this situation totally dominated social and economic processes. At the same time, the states encompassed by these borders started to function as more closely unified systems. This phenomenon is illustrated by the configuration of the Hungarian settlement system (Figure 4.8).

There is no doubt that this phase of rigid ideology, accompanied by distrust and suspicion between countries and the hiding of inefficient social and economic systems behind closed borders did immense harm to the nations sharing the Carpathian Basin at this time. In particular many possibilities of cross-border cooperation were neglected.

The present day situation and future possibilities

A new situation emerged at the end of the 1980s and beginning of the 1990s in all of Eastern Europe including the Carpathian basin. Since that time the influence of a largely independent Ukraine has replaced that of the collapsed Soviet Union. Through the independence of Slovakia another state (besides Hungary) was created with its whole territory in the Carpathian basin. In the south the collapse of Yugoslavia has created three states with problematic relations, bordering Hungary. The political system of all of the above mentioned states has been changed. The removal of the iron curtain and the fact that Austria became a member of the European Union also influenced the situation in the Carpathian basin.

In general the recent changes of political regime have created a more favourable situation for regional cooperation in the Carpathian basin, though much more was hoped for in this respect and inevitably a number of difficulties, weaknesses and suspicions have come to the surface. It is becoming increasingly apparent that development along broadly western European lines will probably affect the whole continent but that the full incorporation of East-Mid Europe will be a long and difficult process.

Hungary, with its central position in the Carpathian basin, and its relatively open economy, is interested in cooperation at all levels including cross-border regional cooperation, as well as international regional cooperation. Attractions across borders are international, and often bring mutual benefits. Emphasizing this latter point is necessary because it reassures those who are fearful of change. Dissolving the highly centralized structures and processes of former regimes would solve many regional problems (poor facilities, unemployment, transport problems) and would bring many social benefits. Good intent, trust and ambitions for mutual advantages would be important requirements for regional cooperation and development, but local geographical coherence would provide the spatial foundations (Figure 4.9).

Figure 4.9   Regional structure of Hungary and border region attractions

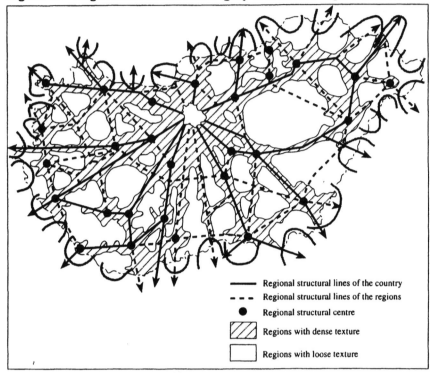

In respect of larger-scaled regional cooperation, four main strategically important connections could be considered, of which two or three are overwhelmingly important. These territories touch all the states of the Carpathian basin and possess many characteristics (Figure 4.10).

The most important strategic connection is the western one, functioning through two international regional relationships. The first one with Vienna is the most important linkage for the whole basin, moreover it has an indirect effect on regions over the whole Carpathian basin. The other international regional relationship towards the West is the four bordered Austrian-Hungarian-Slovakian-Croatian area, which is still under developed, but in the future it will have a great importance.

The most important regional relationship towards the north via Bratislava, is common with the Vienna one. The second one is the attraction of the Hungarian capital to the Mid-Slovakian regions, while the third is focused upon the centres of Miskolc and Kosice.

Figure 4.10   International regional-structural relation system of Hungary

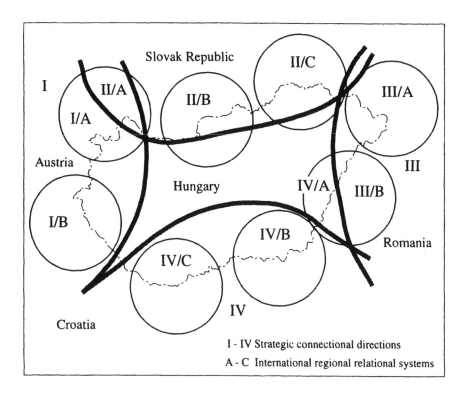

The most important regional connection in the East is the Slovakian-Ukrainian-Romanian-Hungarian border land, the meeting point of Záhony-Csap-Ágcsernyô. This is an existing area of the Carpathian Euroregion on paper, with great potential for future cooperation.   The other Eastern international regional relationship is structured around the region of the Great Plain on the Hungarian-Romanian border.

The first regional relationship with southern connections is common with the above mentioned, while the second one is the Romanian-Serbian and the third involves the cooperation of the areas along the Hungarian-Croatian-Serbian borders, an important Balkan relationship.

These international regional relationships play a very important role in the development of the social and economic potential of the small border regions, and in improving the living conditions of those who live there.   Overall; improved systems of improved regional organization and cooperation within the new political framework could bring significant development energy to many parts of East Central Europe, especially those inside the Carpathian basin.

References

Blahó, A., Palánkai, T., Rostoványi, Z. (1989), *Integrációs Rendszerek a Világgazdaságban*, KJK: Budapest.

Erdösi, F. (1990), 'A Regionális Fejlödés új Mozgatórugója?', *Közgazdasági Szemle*, XXXVII, pp.222-32.

Gazdag, F. (1992), *Európai Integrációs Intézmények*, KJK: Budapest.

Golobics, P. (1994), 'A Nemzetköl Regionális Együttmüködés Közigazgatási Vonatkozásai', in Tóth, J., Mátrai, M. and Székesfehérvár-Pécs (eds), *A Középszintü Közigazgatás Reformja Magyarorzágon 2 Kötet*, pp.39-43.

Golobics, P. (1995), 'A Határmenti Térségek Városainak Szerepe a Regionális Együttmüködésben', *Közlemények a JPTE Általános Társadalomföldrajzi és Urbanisztikai Tanszékéről*, 3, sz. Pécs, p.33.

Golobics, P. and Tóth, J. (1992), 'A Gazdasági Együttmüködés Lehetöségei és Korlátai a Kárpátok-Tisza Régióban', *Regionális Politikák és Fejlesztési Stratégiák az Alpok-Adria Térségben*, Keszthely, pp.23-32.

Golobics, P. and Tóth, J. (1996), 'Ekonomicseszkij Roszt', *Mezsdunarodnoje Szotrudnyicsesztvo (Vvengrija na Poroge XXI-vo Veka)*, Ulan Bator: Mongolia, p.20.

Horváth, Gy. (1992), 'Az Európai Integráció és Területi Együttmüködések Hatása a Piacgazdaságok Regionális Politikájára. *Tér és Társadalom*, 4, pp.51-68.

Horváth, Gy. (1992), *A Közép-európai Integrági Indikátorai*, MTA RKK, Pécs (manuscript).

Inotai, A. (1994), 'Az új Regionalizmus a Világgazdaságban', *Külgazaság*, 1, pp.28-44.

Perczel, Gy. and Tóth, J. (eds) (1994), *Magyarország Társadalmi-gazasági Földrajza*, Elte: Budapest.

Rechnitzer, J. (1992), 'Regionális Együttmüködések és Tapasztalataik', *Az Osztrák-magyar Határmenti Kapcsolatok és az Alpok-Adria Munkaközösség*, MTA RKK, Pécs, (manuscript).

Spath, L. (1991), *1992 Európa Álma KJK*, Budapest.

Süli-Zakar, L. (1994), 'Regionalizmus és Régió', in Tóth, J., Mátrai, M. and Székesfehérvár-Pécs (eds) *A Középszintü Közigazgatás Reformja Magyarországon*, Vol.2, pp.14-22.

Tóth, J. (1992), *Magyarország Illeszkedése a Régiók Európájában*, MTA RKK, Pécs (manuscript).

Tóth, J., Trócsányi, A. and Wilhelm, Z. (1996), *Regions and Interregional Relationships in the Carpathian Basin*, Paper to the International Conference, Maribor, p.25.

5 Infrastructure and regional planning: ownership and regulation of energy, water and land in the English West Midlands and Catalonia

Tim Marshall

The issue

This chapter presents an argument about contemporary regional environmental planning in two EU states, one northern, one Mediterranean. The argument draws on a brief set of historical and descriptive elements, presented partly in tables. The purpose is to provide some historical-geographical perspective on two policy sectors:-

1   regional (or, in Spanish terms, territorial) planning

2   three infrastructure sectors, which are particularly important for environmental policy and regional planning - energy, water and land.

The reason for this focus is that much current debate appears to ignore the context and potential of regional planning and the broader importance of recent changes in the three infrastructure sectors. The focus is therefore at a middle level, not examining the details of regional planning or intra-sectoral changes, but also not simply referring to some very broad shifts that are occurring rather generally throughout Europe, such as changing forms of governance, liberalization and internationalization.

In the past infrastructure and regional planning were analysed together, with one task of the regional scientist, economist or geographer being to identify "bottlenecks" impeding regional growth. From at least the 1950s such an approach was dominant in many European countries, and in 'practical politics' it often remains so. Recent examples of such an outlook are provided in Britain by Vickerman (1991), in Spain by Carbonell (1990) and at the EU level by the large scale quantitative study coordinated by Biehl (1986). Evidence of its continuing dominance in state politics can be found in the UK government's competitiveness White Papers (Department of Trade and Industry, 1994; 1995; 1996), in the Spanish government's Plan Director de Infraestructuras -PDI- (Ministerio de Obras Públicas y Transportes, 1993) and, for the EU, in the Trans European Networks in the transport, energy and communications fields (summarized usefully in Bowers, 1995).

In discussions of the implementation of regional planning in its heyday, the control of direct public investment, especially in infrastructure projects, was presented as one of the four chief instruments, the others being the physical control of land use, directed migration and financial location incentives (Friedmann, 1972). In rich countries all of these instruments remain in use except directed migration, but their nature has been changed dramatically. One dimension of such change has been that the bracketing together of regional planning and infrastructure development has been challenged in the 1990s by green thinking, which suggests that economic growth should not be the only, or necessarily the main, objective of regional policy and planning. In some countries (Sweden, Germany, Netherlands) these ideas have been affecting regional planning since the 1970s. In the 1990s Spain, the UK and the EU all began to refer to environmental goals as significant, so that all the policy documents quoted above make secondary reference to environmental safeguards. The 'overcoming infrastructure bottlenecks' mentality is no longer therefore completely dominant and a new set of issues is emerging, where growth of supply is not necessarily the prime consideration.

Evidently this more environmentally conscious approach is one contemporary trend capable of being taken further. For example the work by Friends of the Earth Europe (1995) is promoting a more radical new agenda, for a 'Sustainable Europe', based on the concept of 'environmental space', with country studies published in 1996-1997. Kunzmann considered a 'Europe of Sustainable Regions' amongst his scenarios for future European spatial development, whereby a region of say five to eight million people might aim to make locally 80 per cent of all its consumer products (Kunzmann, 1996). But this is clearly presented as one of the less likely outcomes.

The relations between infrastructure and regional planning are explored here in two contexts, Spain and the UK. These have the advantage of giving, at a very broad level, some idea of north-south variation within Western Europe. But the more important purpose is to give examples of analysis of specific cases, in order to stimulate discussion and encourage better grounded conclusions about contemporary economic-environmental dynamics in European regions.

The argument

The argument will first be presented schematically. In essence it considers that any attempt to develop a radical social and ecological agenda for these two regions (and no doubt many others) would come up against strong obstacles due to emerging, or existing, shifts in key infrastructural sectors. In other words one of the main instruments for the implementation of regional planning is being weakened. This argument is explored through the examples of the English West Midlands and Catalonia (see Marshall, 1997a, for a related earlier discussion). Their suitability as cases may be contested. Both regions are economically and ecologically wide open systems and the West Midlands has relatively little political autonomy at regional or city level. But it can be argued that there are sufficient grounds for examination of this spatial level:-

1   *Some* ecological aspects of land, water and energy, are operative at or below this level, even if very many (especially in energy terms) are not.

2   There is *some* political capacity at this level, whether with real political and cultural legitimacy (Catalonia) or through unstable, but possibly emergent, groupings of municipalities, partnerships and central government agencies (West Midlands).

3   Regional planning *is* now undertaken, which was not the case before 1980 in Catalonia and between 1979 and 1990 in the West Midlands; that planning claims to take increasing note of 'sustainability' issues (more so in the West Midlands).

4   It may be expected that radical green concepts which have emerged in the last decade, such as 'ecological footprints' (Rees, 1992) or 'environmental space' (van Brakel and Buitenkamp, 1992) will present regional actors with a much more conflictual agenda.

No historical dimension will be given here, but it is important to remember that shifts in planning and in controls over infrastructure during this century have been large and continuous. Whilst this is obvious, it would appear that, when people think about the future from the apparently fixed present, they often forget how extraordinarily mutable the matters they are viewing have been (Marshall, 1997b). Nothing has stood still. Furthermore, 'unintended' interrelationships may often have been as important as those intended by powerful actors, blocking an interpretation of a strongly teleological kind.

Before considering the three sectors, a brief picture of contemporary regional planning will be given. The overall message here is that progress has been so far very weak, although this is less the case in the West Midlands than in Catalonia.

Present day regional planning

Catalonia

The Pla Territorial General (PTG) was approved by the Catalan government in 1995, after 15 years preparation (Marshall, 1995). It is a broadly 'developmentalist' plan, to incorporate 25 per cent more people by 2026, and more economic activity, as well as to shift the distribution of these elements away from the Barcelona metropolitan region. For this it sees the need for more energy resources (primarily North African gas), more water resources (mainly brought from the River Ebro) and more urbanized land. Late modifications introduced the concept of sustainable development into the Plan, but without altering the main policies. Some limited emphasis is given to water and energy conservation and to the development of renewable energies, but this is secondary. Land protection is also emphasized, but primarily in more mountainous areas; the relatively firmly protected zones cover only about 24 per cent of the territory.

West Midlands

Regional Guidance was published by the British government in 1995 (Department of Environment, 1995), using much of the advice of the West Midlands Regional Forum of Local Authorities, which had worked on a strategy from 1990 to 1993 (West Midlands Regional Forum of Local Authorities, 1993). This was the first region-wide governmental 'plan' since the 1970s. Although much shorter than the PTG (about 70 pages versus, in total, over 650 pages), the guidance is perhaps almost as significant in land use planning terms, representing an important constraint on the actions of statutory planning authorities (counties and districts), as well as forming some input to the policy making of central government within the region. This was reorganized in 1996, forming the Government Office for the West Midlands, including some of the key economic-environmental ministries (Trade and Industry, Environment, Transport, Education and Employment).

The Guidance puts strong emphasis on sustainable development, and is at least verbally insistent on the importance of conservation of all natural resources including land, water and energy. Its goals are therefore not so developmentalist as those of the PTG. But there is little evidence that the Guidance can provide

56

ways of achieving these goals, and it often appears that, where specific strategic measures are indicated, these go against these goals. For example broad support is given for land release for 'premium employment sites' and 'major investment sites' on the edge of the West Midlands conurbation (paras 7.9 to 7.20). Major road schemes are supported, but there is little sign of planning for major public transport investment. The goals of economic growth and sustainability are simultaneously present but the relations between them, or the nature of trade offs, are not made clear. This inevitably causes a tension in the guidance, which undermines any radical concern for sustainability. As in the Generalitat (Catalan government), the prime goal of the Regional Office would appear to be competitiveness (Mawson and Spencer, 1995), to be achieved primarily through non-interventionist neoliberal means. A document on 'regional competitiveness' confirms this judgement, although this concept is considered by government policy to be compatible with sustainability (Government Office for the West Midlands, 1996). The following sections examine the chosen sectors in each region. They reveal a critical absence of usable instruments to achieve progress towards the goal of sustainability, a goal which is clearly enunciated in the West Midlands and is beginning to be explored in Catalonia.

Recent shifts and prospects in each sector

Water

There are some parallels between recent changes in the two regions, but given the significantly different starting points, the implications for regional planning do vary considerably. (Table 5.1 gives a brief summary).

One common factor is the still major role for state regulation, in relation to pricing, investment and quality standards. This regulation is shared in Britain between central government agencies (Department of Environment, the Environment Agency and the Office of Water Regulation - Ofwat) and the EU, and in Catalonia between regional, central and EU levels. The West Midlands governmental agencies have little control over the region's water resources, and in Catalonia the Generalitat's control is only partial; it has significant powers over pricing and investment, but must defer to Madrid in key inter-basin issues and overall standards (see MOPT, 1993 and Institut Català d'Energia 1992; NRA, 1994a; 1994b).

It is far from easy to judge which region experiences the stronger state regulation, with the situation being quite fluid, especially in Britain. But it would probably be fair to say that in 1996, there remained in both cases significant regulatory powers, which *could* be harnessed to a broader range of social and ecological ends than those sought at present. So far, this regulation has been addressing the emerging issues of demand management and supply efficiency

Table 5.1
Ownership and regulation of water

REGION	OWNERSHIP	REGULATION
Catalonia	Aigues de Barcelona (Lyonnaise des Eaux, La Caixa and electricity companies) in metropolitan area, and smaller private and municipal companies elsewhere.	EU, central government hydrological plan (PHN), and Catalan government's environmental and public works agencies and departments.
West Midlands	Severn Trent, near monopoly private company.	Environment Agency (previously National Rivers Authority), Office of Water Services (OFWAT), Department of Environment (DOE), EU - all these primarily supraregional.

especially in Britain, with water protection being a high priority in the NRA's liaison with the planning system (Guy and Marvin, 1995a). Nevertheless Regional Guidance does identify a potential supply shortfall by 2021, if no further water resources are developed (Department of the Environment, 1995, para 12.18). The continuing force of the water industry's 'supply mentality' is therefore evident in the Guidance.

In Catalonia the main current concerns are prices to consumers and the maintenance of adequate supplies. The dates forecast for supply shortfalls vary considerably depending upon the part of Catalonia, but the new plan for Catalonia's internal basins, submitted to Madrid in September 1995, argued that a large overall deficit would develop by 2012 in the metropolitan region.

Water industry ownership and control may be seen to be converging in the two regions. Privatization of the Severn Trent Regional Water Authority in 1989 (covering an area larger than the West Midlands) created a powerful and dominant water supplier in the region, with only one other company controlling a small area. So far Severn Trent has not been subject to take over bids, but change to multinational ownership is likely within the next few years, because of the attractiveness of the solid, state regulated monopoly markets represented by English and Welsh water companies. In Catalonia, Aguas de Barcelona has a degree of dominance in the core Barcelona metropolitan region, and has been taking over smaller municipal and private companies elsewhere in Catalonia and Spain. It is itself 43 per cent controlled by Lyonnaise des Eaux, France's second largest water company, and La Caixa (Catalonia's largest bank), with further large holdings, totalling 24 per cent, by Endesa and Iberdrola (electricity firms) and Banco Bilbao Vizcaya.

It is reasonable to expect that this process of concentration and internationalization will continue, creating very large international companies in water supply, water treatment and waste management. This trend is being boosted by EU policies, which somewhat lessen central state regulation and impose technologically demanding standards, often not easily met by smaller water companies.

What is emerging then in the water sector is apparently a more concentrated and powerful industrial force, which may increasingly be able to bargain with state regulators, rather than accept their edicts. This is likely to lessen the scope for water planning, whether on a central or regional basis, if that planning seeks goals not consistent with long term profit maximization. Powerful socio-ecological and political movements *might* be able to control this tendency but the power necessary to achieve this would have to be that much greater than would otherwise have been the case (or than say 10 or 20 years ago), given the combination of forces present in central government, large companies and the EU.

Given these circumstances, regional planners may have little temptation to expand their involvement in the planning of one of the key bases of each region. Or, alternatively, they may be encouraged to take the path of least resistance, by matching their plans for water supply more closely to the desires of the private-sector companies involved.

Energy

Developments in the energy sector have been highly contradictory in recent years, in both regions, but in both cases have resulted in major restructuring. Evidence for this has been abundant in the press, but also more formally in the work of Guy and Marvin (1995b), Graham and Marvin (1995), McGowan (1993) and McGowan and Thomas (1992). A summary is in Table 5.2.

In Spain energy industry ownership and control has been rearranged between central government and large gas, oil and electricity companies, particularly since the approval of the last national energy plan (PEN) in 1991. This has resulted in the gradual disposal of state assets in these sectors, particularly in Repsol (oil) and Enagas, although one electricity giant, Endesa, remained 65 per cent state owned in 1996. In gas the private virtual monopolist is now Gas Natural. This has been built round a Catalan core company, but is now dominated by the

Table 5.2
Ownership and regulation of energy

REGION	OWNERSHIP	REGULATION
Catalonia	Endesa (all Spain primarily state owned electricity company), Gas Natural and Repsol (mainly privately owned gas and oil companies, with large Catalan participation, particularly by La Caixa, plus other Spanish banks).	National Energy Plan (PEN), and EU, which has some price, investment and environmental controls, particularly via cross frontier projects, e.g. pipeline link to Algeria.
West Midlands	Private production and distribution companies, regional electricity company at present based in region.	Environment Agency (previously Her Majesty's Inspectorate of Pollution), Department of Trade and Industry (DTI), Offices of Gas and Electricity Regulation (OFFER/OFGAS), EU - complex and mainly weak structure.

mainly private Repsol (45 per cent) and La Caixa (25 per cent). In electricity Endesa and three or four private firms (including one as large as Endesa, Iberdrola) have divided up the generating and distribution market for all of Spain, with transmission still 51 per cent state owned. The conservative government elected in March 1996 intended to dispose of all industrial sector state assets within two years, and so the remaining holdings in the energy field are likely to be privatized quite quickly.

The effect in Spain has been, therefore, to create a few powerful and mainly privately owned Spanish energy corporations. These control distribution and sales of oil and gas, and produce electricity, currently derived mainly from nuclear, oil, hydro and coal, but with the PEN aiming to increase gas based generation. Oil and gas production has been mainly globally controlled for many years, but the above corporations remain primarily Spanish owned at present, now with little regional differentiation. They are expected to compete and diversify on the world market. They are closely linked to the largest Spanish banks. Within the Spanish economy two very powerful financial - industrial complexes are seen to be emerging, one centred on the Banco Santander, the Banco Central Hispanico and Endesa, the other on Banco Bilbao Vizcaya, La Caixa and Repsol/Gas Natural (Burns, 1996). EU liberalization of the gas and electricity industries (with third party access and other key measures) has been under discussion as part of the Single Energy Market initiative since 1988, and is likely to be finally agreed in the near future. This will allow the break up of national monopolies, and probably lead to the same pattern of international or global ownership as is emerging in the water and waste sectors. The movement of regulation to the EU level may be logically following, although this might be expected to be much less strong than the central state regulation of most of the last 50 years.

For the present therefore Catalonia faces increasingly powerful privatized energy industries operating across the whole of Spain. State regulation remains very significant, primarily based in Madrid. The Generalitat's main involvement is with the support of energy conservation and renewable energies, but this is on a relatively limited scale, given small budgets. Madrid's power in setting state regulation via the regular PENs may be expected to decline, as the energy corporations flex their muscles, and especially if they become internationalized (perhaps along with the banking system). For the present there is a kind of interregnum, but again there seems little temptation for regional planners to imagine that they could control significantly any levers over regional energy production or use. The idea that Catalonia might move 'backwards' to its relatively sound energy base of 1950 (90 per cent hydro power), as well as forwards to other renewables, and away from its overwhelming dependence on gas, oil and nuclear power, looks at present an unlikely prospect.

In the West Midlands the similarity is strongest in industry ownership and control - the presence of large monopolistic or duopolistic energy corporations, created in the privatizations of oil, gas, electricity, coal and (partially) nuclear power. Some British regional distribution companies have already been taken over by foreign companies or by one of the major electricity producers. The West Midlands is fuelled primarily by the two major national electricity producers, the regional electricity company and the so far dominant, but threatened (and now divided), British Gas. If the regional electricity company is taken over, the localized dimension of the energy sector will become minimal - as it was in the days of the state owned corporations.

The British government avoids all suggestion of a policy of state energy regulation, although through its Offices of Electricity and Gas Regulation, and other bodies concerned with coal, oil and nuclear exploitation, its role in market and price regulation is very significant. No regional role is present in this regulation, and it is understandable that Regional Guidance devotes only one, non-committal page to energy, all on renewable energies. Other sections, on air quality and travel minimization, do of course have significant energy implications, however. Again the claims to sustainability clash with the limited scope for any regionally oriented public regulation. More broadly the possibility of strong state regulation appears to be declining given the emerging power of energy corporations and the likely shifting of regulatory power (however restrained) to the EU. A radically different energy policy for parts of Britain would have to face up to these emerging difficulties, just as in Spain.

Land

It is interesting to look at regulation of land at the same time as that of energy and water, despite their very different relations to ecology, economy and society, because this brings out the highly differential control exercized over natural and social resources (see Table 5.3).

Table 5.3
Ownership and regulation of land

REGION	OWNERSHIP	REGULATION
Catalonia	Mainly private, but some state and municipal ownership remaining, and Generalitat agency (INCASOL) leads market in residential and industrial development outside major metropolitan areas.	Planning and control systems strongest at municipal levels, but powerful regional guidance of roads investment. Regional plan (PTG) relatively weak.
West Midlands	Virtually all private.	Fairly strong local and regional guidance systems, with roads investment at state level.

It is evident that influence over land use change lies at the core of both the Catalonia PTG and West Midlands Regional Guidance. Both seek above all to influence the location of urbanization, over varying periods, through their effect on the statutory land use plans in each country. In this sense the control over the resource is much more direct, based on legislative systems built up in each country over the last 50 years or more. That system has been made effective in Spain over the last 15 years, and periodically weakened in Britain over the same period. In both cases however an apparently powerful system of control was in place in each country in 1996, and the regional documents form at least potentially significant elements of these systems. Nevertheless the system is still used essentially for overcoming physical bottlenecks, reducing the friction of space, particularly in allocating land for building transport links and facilities including motorways, car parks and high speed train corridors. The influence from the EU on the land regulation systems so far is also slight.

In both countries land ownership is now predominantly private, with the dismemberment of the public sector landholdings being very far advanced in Britain, to the extent that probably only in certain residual forms (military land, some forest land) does it remain significant. Semi-public forms, like the National Trust, may now matter more. Most development land, at any rate, is likely to be passing through private markets; this has been the case in housing, industry, retailing and leisure in most of Britain for much of the last two decades.

In Catalonia public land disposal has also advanced greatly since the end of Francoism, but a number of state and municipal agencies remain significant landholders, with influence particularly over key development zones. These include port and airport authorities, some larger municipalities, and the railway corporation. Gradually these holdings are being reduced, under various political and economic pressures, and such pressures will intensify, with conservative control of all levels of government completed in 1996.

However, in Catalonia the role of INCASOL, the Generalitat's land assembly agency, is of central importance in the housing and industrial sectors. This has changed the dynamic of urbanization significantly in those areas where it has been most active (above all outside the Barcelona metropolitan region). Within the Barcelona metropolitan region other, municipal or municipal alliance, bodies have had a similar role in land assembly. How far this role will be able to continue with conservative government in Madrid is uncertain.

Influence over the use of land is therefore likely to depend, during the next 10 years or so at any rate, on mechanisms of public regulation, rather than ownership. This is already so in Britain, and may be increasingly the case in Catalonia. In that sense the regulation mechanisms face, or are likely to face, the same powerful types of actors as in the water and energy fields.

In principle however, the regulation mechanisms are much more practiced and effective in this field than they are in respect of water or energy, and established land use control systems are backed up by a range of further forces. In Britain these include publicly funded nature or rural protection agencies (e.g. English Nature, Countryside Commission) and powerful rurally based (i.e. development zone based) protection movements such as the Council for the Preservation of Rural England. In Catalonia these forces are normally not so strong, but are mobilized in some of the key zones, for example much of the coast and some agricultural districts, by other sorts of actors, including second home owners, farming interests and green organizations based in the cities.

In this case therefore there are strong mechanisms, above all, the statutory planning systems, backed up by a variety of place-based social forces, which exert some influence over changes in this most visible resource. This contrasts sharply with the scope for control over water and energy. Arguably water and energy would be of equal importance in the formulation of long term ecological and social strategies for regions, but they lack both regionally oriented control and regulation systems, and (at present) social interests which might push for, and against, such systems. No doubt this is in part due to the presence of direct public control over water and energy for much of this century in both countries, which, however unevenly, has provided some element of social influence over water and energy change. Now that this control has been transformed and, it appears at present, diluted, quite new issues are raized.

Conclusions

It is worthwhile highlighting some underlying emphases in this account. The aim has been to stimulate consideration of the current state of control of the three resource sectors and the implications for future regional planning approaches. This looks forward to a form of regional planning which takes into account biophysical bases of livelihood, as well as social dimensions. It envisages that in the next planning cycle, water and energy issues will not be seen primarily as 'bottlenecks' to regional development, but as long term foundations for regional livelihoods.

The main mechanisms which have been picked out in the account are:-

1   the changing forms of state regulation, with the role of the EU particularly significant for water and energy, and an emerging arena of regulation of prices, investment and quality standards, so far little related to physical or territorial planning;

2   the privatization and possible internationalization of industries in these sectors, creating a balance of power different from that present in the earlier histories of regional and urban planning;

3   the variable force of politically relevant actors, with land issues being

generally more politicized, at least in a more spatially targeted form, than those concerning water and energy.

Of course privatization and liberalization have effects other than those discussed here, for example in changing the balance of the regulatory structure and in altering the forms of openness and secrecy within the affected sectors. These further effects partly condition the terms on which public influence may be brought to bear on the new private actors, but they do not change the transformed overall structure of each sector.

It has been suggested that the evolution of the three mechanisms in state, capital and society outlined above will be of central importance for any socially and ecologically responsive regional planning in the future. This planning will have to tackle questions of 'geographical displacement' (Haughton, 1995), and move towards forms of 'quasi-autarkic' (more self reliant, mutually compensating) cities and regions (Hunter and Haughton, 1996). From this perspective, current policies at EU and central state level, on the deregulation and liberalization of key environmentally related sectors, are likely to need massive adjustment if regional planning is to be able to achieve its potential as a key level of planning in the coming decades. One element of such adjustment would be the creation of a framework for demand side management; this already exists for land to varying degrees, and is much discussed in the fields of energy and water (Graham and Marvin, 1995). This could be one tool for the further promotion of renewable energies (Boyle, 1996).

It is not clear, however, whether the pressure for such consideration of radical policy change could come from within tensions surrounding the water, energy and land 'sectors', or whether a more central, political source, considering new regional (or national) economic models, is more plausible. If one of the future focusses is to be on demand management of water, energy and land use, then the regional industrial structure will be one of the main keys, with many regions wishing to avoid, in future, industries which are heavy water and energy consumers. But this may leave the economic base of regions very dependent on sectors with equally difficult, if different, problems, for example tourism, finance, retailing. This line of thinking then, would imply that 'competitiveness', in all its guises, does remain at the core of any consideration of water, energy and land futures.

This chapter suggests that these considerations need to progress *together*. This conclusion, at any rate, would not create surprises for many politicians and officials in the Generalitat or the West Midlands Regional Forum. But it does seem to escape many academics, in various fields, and is rarely confronted seriously by politicians at higher levels. Debates on sustainability and regional competitiveness need to be joined and reframed, and powerful new kinds of public control (spatial as well as non spatial) need to be fashioned. Otherwise, the scope for a radical socio-ecological regional planning, in these regions as elsewhere, will be increasingly constrained by the decisions of internationalized infrastructure corporations. Given the very long time scales involved in infrastructure investment and therefore the way that futures can be hedged around for decades, this is an issue that needs urgent and serious consideration by environmental policy makers.

References

Biehl, D. (1986), *The Contribution of Infrastructure to Regional Development*, Final Report, Infratructure Study Group, CEC: Brussels.

Bowers, C. (1995), *Ten Questions on TENs*, European Federation for Transport and Environment: Brussels.

Boyle, G. (ed.), (1996), *Renewable Energy*, Oxford University Press: Oxford.

Burns, T. (1996), 'Two Duellists Take Guard', *Financial Times*, supplement on Spain, p.II, June 24 1996.

Carbonell, A. (1990), *Las Infraestructuras en España: Carencias y Soluciones*, Instituto de Estudios Economicos: Madrid.

Department of the Environment (1995), *Regional Planning Guidance for the West Midlands Region*, HMSO: London.

Department of Trade and Industry (1994), *Competitiveness, Helping Businesses to Win*, HMSO: London.

Department of Trade and Industry (1995), *Competitiveness, Forging Ahead*, HMSO: London.

Department of Trade and Industry (1996), *Competitiveness*, HMSO: London.

Friedmann, J. (1972), 'Implementation', *International Social Development Review*, No 4, pp. 95-105.

Friends of the Earth Europe, (1995), *Towards Sustainable Europe: A Summary*, Friends of the Earth Netherlands: Amsterdam.

Government Office for the West Midlands (1996), *Working to Win. A Framework for Competitiveness in the West Midlands*, GOWM: Birmingham.

Graham, S. and Marvin, S. (1995), 'More Than Ducts and Wires: Post-fordism, Cities and Utility Networks', in Healey, P. et al. eds, *Managing Cities*, Belhaven: London.

Guy, S. and Marvin, S. (1995a), *Planning for Water: Space, Time and the Social Organization of Natural Resources*, Working Paper 55, Department of Town and Country Planning, University of Newcastle upon Tyne.

Guy, S. and Marvin, S. (1995b), 'Changing Logics of Infrastructure Management', paper for British Sociological Association, University of Leicester, April 1995.

Haughton, G. (1995), 'Regional Resource Management, Sustainable Development and Geographical Equity', paper presented at the ESRC Sustainable Cities seminar, 5/9/95, Manchester Metropolitan University.

Hunter, C. and Haughton, G. (1996), 'Sustainable Urban Development and Fresh-water Resource Management', *Sustainable Urban Development Working Papers*, 2, Leeds Metropolitan University/University of Aberdeen.

Institut Català d'Energia (1992), *Quaderns de Competitivitat. L'Energia i L'Aigua*, Generalitat de Catalunya, Departament d'Indústria i Energia, Direcció General d'Indústria.

Kunzmann, K. (1996), 'Euro-megalopolis or Themepark Europe?', *International Planning Studies*, Vol 1, No 2, pp 143-163.

McGowan, F. (1993), *The Struggle for Power in Europe*, RIIA: London.

McGowan, F. and Thomas, S. (1992), *Electricity in Europe. Inside the Utilities*, FT Business Information: London.

Marshall, T. (1995), 'Regional Planning in Catalonia', *European Planning Studies*, Vol 3, No 1, pp. 25-45.

Marshall, T. (1997a), 'Dimensions of Sustainable Development and Scales of Policy Making: Birmingham/West Midlands and Barcelona/Catalonia', in Baker, S., Kousis, M. and Young, S., *Sustainable Development: Theory, Policy and Practice in the European Union*, Routledge: London.

Marshall, T. (1997b), 'Futures, Foresight and Forward Looks. Reflections on the Use of Prospective Thinking for Transport and Planning Strategies', *Town Planning Review*, Vol.68, No.1, pp.31-53.

Marvin, S. and Cornford, J. (1993), 'Regional Policy Implications of Utility Regionalization', *Regional Policy*, Vol 27, No 2, pp. 159-65.

Mawson, J. and Spencer, K. (1995), 'The Government Offices for the English Regions', in Hardy, S. Hebbert, M. and Malbon, B. (eds.), *Region Building*, Regional Studies Association: London.

Ministerio de Obras Públicas y Transportes (1993), *Plan Director de Infraestructuras 1993-2007*, MOPT: Madrid.

National Rivers Authorities (1994a), *Water, Nature's Precious Resource. An Environmentally Sustainable Water Resource Development Strategy for England and Wales*, NRA: Bristol.

National Rivers Authority (1994b), *Guidance Notes for Local Planning Authorities on the Methods of Protecting the Water Environment through Development Plans*, NRA: Bristol.

Rees, W. (1992), 'Ecological Footprints and Appropriated Carrying Capacity: What Urban Economics Leaves Out', *Environment and Urbanization*, Vol 4, No 2, pp. 121-30.

Stern, J. (1992), *Third Party Access in European Gas Utilities*, RIIA: London.

van Brakel, M. and Buitenkamp, M. (1992), *Action Plan, Sustainable Netherlands*, Friends of the Earth Netherlands: Amsterdam.

Part Two
THEMES AND ISSUES

6 Integration of land and water management in England

Nigel Watson

Introduction

During the last ten years there have been numerous calls for integrated management of land and water resources (for example, Lundquist et al., 1985; OECD 1989; Downs et al., 1991; Newson 1992; Deyle 1995; Margerum and Born 1995; Watson et al., 1996). While integrated approaches have been practised in the water resources sector for some time, it is now widely recognized that a broader interpretation is required to address problems arising from interactions among land and water systems. Examples of such problems can be found in most if not all parts of the world, and include erosion, nitrate and pesticide pollution, flooding, and drought.

Despite strong support among academics and practitioners, implementation of integrated management approaches for land and water has been slow. One reason for this is that little consensus exists regarding what the term 'integrated management' actually means (Born and Sonzogni, 1995). For example, Downs et al., (1991) identified thirty-six different terms in the research literature which are used to describe the aims, geographical focus and functions of integrated management. According to Mitchell (1986, p.13), integrated resource management "is the sharing and coordination of the values and inputs of a broad range of agencies, publics, and other interests when conceiving, designing and implementing policies, programs or projects". Thus, integration implies joint decision making with the aim of balancing diverse public and private interests at different levels of decision making. A second reason for slow implementation is that there are often numerous institutional barriers and constraints which prevent the development of coordinated or linked arrangements for decision making. For clarity, the terms institutions and institutional arrangements as used here refer to the structures, processes and policy approaches for making public decisions and for influencing the behaviour of individuals, groups and firms (after Mann, 1983, p.116). Examples of institutional barriers and constraints include overlapping agency responsibilities, fragmented administrative structures, weak legislation, inadequate financial provision, limited public participation and entrenched organizational cultures (Ingram et al., 1984; Born and Rumery, 1989; Mitchell, 1990).

In this chapter, the effectiveness of the institutional arrangements for integrated management of land and water resources in England is examined. Attention is focused on the management of nitrate pollution in order to illustrate how institutional reform has led to improved integration of the land and water sectors. The chapter begins with an account of the nature and significance of the nitrate problem. This is followed by an evaluation of institutional responses for nitrate pollution. Five evaluative criteria are used: coordination; public participation; mix of strategies; adaptive capacity; and equity, efficiency and effectiveness. Evidence was collected through interviews and a questionnaire survey which included representatives for agricultural, environmental, government, and water management interests. Information was gathered from national organizations and representatives for agencies and groups in the Anglian, Severn-Trent, and Yorkshire water management regions. These three regions were selected because each has a significant nitrate pollution problem. In the final section, conclusions from the study are discussed and recommendations to improve the management of the nitrate problem are offered.

Nitrate in water: problems and key issues

The need for integrated management of land and water is clearly illustrated by the problem of nitrate pollution. Changes in agricultural practices during the last fifty years have resulted in nitrate contamination of a significant number of surface and ground water sources in England. One hundred water supplies serving 2.25 million people exceeded the 50mg/litre standard for nitrate between January 1994 and June 1985 (House of Lords, 1989). By 1990, the number of people receiving such supplies had increased to 5.3 million (Drinking Water Inspectorate, 1992). Nitrate pollution is an excellent example of a resource "meta-problem" (Trist, 1983), "wicked-problem" (Dorcey, 1986), or "mess" (Ackoff, 1974). These types of problems are characterized by complexity, uncertainty and conflict.

The nitrate problem has developed because of complex linkages among bio-physical, economic and social systems. Since the late 1940s, legislation and government policy in the UK have been directed towards improving the efficiency and self-sufficiency of food production. While the economic transformation of agriculture was a remarkable success, a range of environmental problems emerged in rural areas as a result of intensification of farming. Only in the last five to ten years have UK and EU policy makers attempted to strike a more even balance among the economic, social and environmental objectives for agriculture. The process of nitrate leaching is itself highly complex. There was a substantial increase in the use of nitrate fertilizers in the UK between 1950 and 1985 (Figure 6.1). However, research has shown that use of nitrate fertilizer may not be the main cause of the problem (Addiscott, 1988, Addiscott and Powlson, 1989). A more likely explanation is that nitrate from fertilizer is incorporated within the soil organic matter, and is subsequently leached from the soil several months or years later. Release of nitrate from the soil can be triggered by a number of factors, including changes in temperature and rainfall, ploughing of grassland, or changes in crop types and rotations.

Figure 6.1 UK consumption of inorganic fertilizers

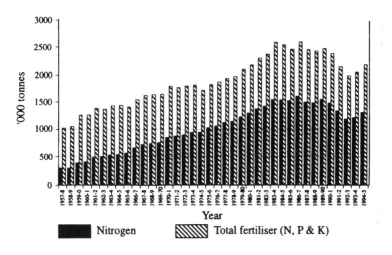

The public health and environmental risks associated with nitrate pollution are shrouded by many uncertainties. The link between nitrate in drinking water and infantile methaemoglobinaemia (blue-baby syndrome) was first reported by Comly (1945). The condition occurs when consumed nitrate is converted to nitrite. A sufficient concentration of nitrite can inhibit the capacity of red blood cells to carry oxygen, the effects of which can be fatal. Fortunately, the condition is very rare in Western Europe and the last fatal case in the UK occurred in 1974. Nitrate consumption has also been linked with gastric cancer, liver disease and respiratory defects (Magee, 1982; Walters, 1984). However, there are still many uncertainties regarding the effects of nitrate upon human health and there appears to be little agreement among medical experts regarding the risks. There are also many uncertainties regarding the ecological effects of nitrate pollution. For example, nitrate can contribute to the eutrophication of surface waters. However, different views have emerged regarding the relative importance of nitrogen and phosphorous in the eutrophication of inland and coastal waters.

Given the complex and uncertain nature of nitrate pollution it is not surprising that conflicts have arisen among agricultural, environmental, government and water management interests regarding what, if anything, should be done about the problem. The challenge, therefore, is to develop institutional arrangements to cope with complexity, uncertainty and conflict. Institutional responses for the nitrate problem in England are examined in the following sections.

Coordination

In many countries, responsibilities for agriculture and water resources management are divided among numerous government departments, public agencies, and private organizations. In these circumstances, complex problems such as nitrate pollution are unlikely to be resolved unless coordinated policies and programmes are developed. Coordination has been defined as "the process whereby two or more organizations create/or use existing decision rules that have been established to deal collectively with their shared task environment" (Mulford and Rogers, 1982, p.12).

Responsibility for policies to control nitrate pollution in England is shared by two government departments. The Department of the Environment (DoE) controls several policy areas including housing, planning, environmental protection and water resources. The Ministry of Agriculture, Fisheries and Food (MAFF) has an overall aim of fostering an efficient and competitive agricultural sector. However, MAFF was given an additional duty under the 1986 Agriculture Act to balance the interests of agriculture with the conservation of the countryside, and therefore has a general duty to prevent agricultural pollution.

Before 1985, there were a number of conflicts between the DoE and MAFF regarding responsibility and approaches for agricultural pollution control and prevention (Watson, 1992). Coordination and cooperation between the two organizations did improve during the late 1980s due to ad hoc and informal arrangements. However, a formal inter-departmental group was established in 1985 to examine the nitrate problem. The Nitrate Coordination Group (NCG) included representatives for the DoE, MAFF, agriculture, research and development organizations, and the water industry. The NCG did produce some agreement regarding the extent and causes of the problem. Unfortunately, a consensus did not emerge regarding the equitable allocation of the costs. For example, the final report produced by the NCG (DoE, 1986, p.77) included the following statement:

74

The nitrate issue raises difficult and apparently conflicting questions of equity which we were not able to reconcile. There are arguments for and against the agricultural industry, the Government or the Community or any combination of these bearing the costs of resolving the nitrate problem.

The NCG was disbanded in 1986. However, similar groups were used on a less formal basis after this date to examine proposals to deal with the nitrate problem, including Nitrate Sensitive Areas (NSAs). Interviews with agricultural officials and representatives for water management indicated there were varying degrees of coordination at the regional level to deal with nitrate pollution. While there were strong links among technical staff dealing with operational matters, there was relatively little contact among agricultural and water management interests regarding strategic issues. A Conservation and Pollution Group was established in the Severn-Trent region to facilitate coordination among MAFF and National Rivers Authority (NRA) officials. However, similar arrangements were not created in the Yorkshire and Anglian regions where there are also significant nitrate problems.

In the early 1990s, the NRA did introduce catchment management planning (cmp) in order to coordinate uses of water and other resources. The process involved the identification of uses of the water environment, their requirements in terms of water quantity and quality, and options to improve conditions where uses were impaired (NRA, 1993). Catchment management planning was used by the NRA as a means of influencing land use planning decisions (Slater et al, 1994). However, this initiative had little effect upon the nitrate problem because agricultural land use is not subject to control through the land use planning system. The programme of catchment management planning ended during 1996 because the functions of the NRA were transferred to a new Environment Agency (EA) for England and Wales. In the future, the EA will produce Local Environment Agency Plans (LEAPs) which will address air quality and waste disposal, as well as water resources issues.

There were also different views regarding the need for greater cooperation and coordination. Several NRA officials argued that stronger links between land and water management were needed. In contrast, a number of MAFF officials believed that further coordination was unnecessary because water management was not one of their organization's main responsibilities.

Public participation

Opportunities should be provided for public participation in decision making because of the diverse range of land and water interests which are affected by the nitrate problem. Agreement regarding the causes, consequences and responses for the problem is unlikely to be reached if some interests are inadequately represented. At the national level, consultation by government departments provided few opportunities for public participation in the development of policies to deal with nitrate pollution. While interested groups were invited to respond to consultation documents issued by the DoE and MAFF, this approach did not provide opportunities for direct participation in decision making. However, a broad range of interests was able to participate in the 1989 House of Lords inquiry for the proposed European Union (EU) Nitrate Directive. Thirty-four organizations submitted written evidence and twelve additional organizations representing agriculture, environment, government and water management gave oral evidence to the inquiry. As a result of the inquiry, the House of Lords Select Committee on the European Communities concluded that, as drafted, the Nitrate Directive would severely affect farmers in the UK and urged the European Commission to revise the proposals. The Nitrate Directive was subsequently modified to enable the Member States to develop measures which equitably balance the interests of agriculture and water management.

At the regional level, MAFF provided few opportunities for non-agricultural interests to participate in consultation processes. For example, non-statutory Regional Panels were established by MAFF to aid communication between government ministers and the agricultural industry. Members of the Panels are appointed entirely by the Minister for Agriculture, and to date representatives for water management interests have not been included. Within the water sector, statutory Regional Rivers Advisory Committees (RRACs) do provide opportunities for a broad range of interests to participate in decision making. The RRACs include representatives for agriculture as well as local authorities, water suppliers, recreation groups, and industry. Nevertheless, the RRACs were created primarily to deal with the water environment, rather than broader issues such as nitrate pollution which involve the use of land.

Mix of strategies

There is a range of strategies which can be used to deal with the nitrate problem. The effects of the problem can be reduced using curative strategies. In contrast, preventative strategies deal with the causes. Adaptive strategies assume that society will adjust to changes in conditions brought about by the problem. All three types of strategy may involve combinations of regulatory, technological and economic measures. Due to the complex and uncertain nature of nitrate pollution, a mix of strategies which include combinations of measures is likely to be the most effective response.

Before 1985, little action was taken to address the nitrate problem because few supplies exceeded the 100mg/litre limit for nitrate set by the World Health Organization (WHO). However, action was prompted by implementation of the EU Drinking Water Directive in 1985, which set a more stringent limit of

50mg/litre. The water supply organizations responded to the lower limit by adopting curative strategies which involved technological measures such as the development of alternative water sources, blending of polluted and higher quality water, and ion-exchange de-nitrification. There were, however, substantial economic costs associated with these measures. For example, the South Staffordshire Water Company estimated the need for a £15 million cumulative capital expenditure for de-nitrification by the year 2000 in order to secure potable water for 130,000 people.

After 1988, greater emphasis was given to prevention by modifying agricultural practices in areas where water sources are vulnerable to nitrate pollution. Combinations of regulatory and economic measures were implemented. For example, ten pilot Nitrate Sensitive Areas (NSAs) covering 10,500 hectares were established by MAFF in 1989. In each NSA, farmers were able to enter into voluntary agreements which provided financial compensation for changes in land management practices or land use. While farmers opted to change management practices for 87% of the NSA land, land use was changed on only 14% of the land in the scheme. Nine Nitrate Advisory Areas (NAAs) were also created to provide farmers with advice regarding ways of reducing nitrate pollution without limiting agricultural productivity. Prevention was strengthened further in 1991 following the introduction of a statutory Code of Good Agricultural Practice for the Protection of Water. Implementation of the Code has meant that farmers are no longer able to avoid prosecution for pollution offences by arguing in court that they have followed good agricultural practices. Regulations for the use and management of silage and slurry were also strengthened through the introduction of minimum design standards for storage facilities. However, there are still gaps in the regulations because farm waste is not classified as a 'controlled waste' under the 1988 Collection and Disposal of Waste Regulations. In effect, the storage of farm waste is regulated, but its use and disposal to land is not.

Reform of the EU Common Agricultural Policy (CAP) in May 1992 provided further opportunities to prevent nitrate pollution. An additional twenty-two NSAs covering 35,000 hectares were established in England as a result of an agri-environment regulation agreed by the Member States. In addition, progress was made towards the designation of Nitrate Vulnerable Zones (NVZs) as required under the EU Nitrate Directive (91/676). Under this Directive, Member States must designate all areas of land which generate runoff where nitrate concentrations exceed, or are expected to exceed 50mg/litre or where there is evidence of nitrate-induced eutrophication, as Nitrate Vulnerable Zones (NVZs). By 1999, measures must be introduced in each of seventy NVZs in England to limit the application of nitrogen fertilizers and organic manure, to regulate the storage of slurry and silage, and to develop plans for fertilizer use for individual farms (Figure 6.2).

Figure 6.2   Proposed nitrate vulnerable zones

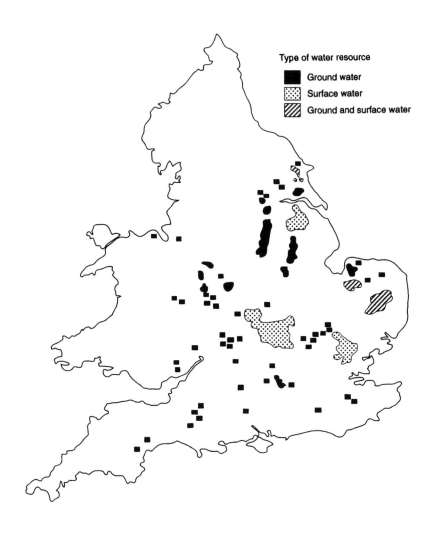

Source: Watson et al, 1996.

Resource meta-problems, messes, and wicked problems require a flexible and adaptive management approach. Advocates for adaptive environmental management have pointed out the dangers and limitations of conventional approaches for resource management. For example, Walters (1986, p.vii) argued:

> We keep running up against questions that only hard experience can answer, and a basic issue becomes whether to use management policies that will deliberately enhance that experience. Such policies would represent a radical departure from the traditional prescriptions about how to deal with uncertainty, namely to proceed with great caution or to act as though there were no uncertainty in hopes that mistakes and opportunities will automatically reveal themselves.

One of the key ideas associated with adaptive environmental management is that options should be kept open for the future by ensuring that policies and programmes are flexible and easily modifiable (Holling, 1978; Walters and Holling 1990; Lee, 1993; Gunderson et al.,1995). Thus, environmental management is viewed as a process of continuous learning whereby approaches are adjusted as the effects of policies become apparent or as conditions change.

Unfortunately, policy makers in England appear to have been reluctant to develop an explicit and deliberately adaptive approach for the nitrate problem. For example, the option of preventing the problem is no longer viable for water sources with nitrate concentrations above 50mg/litre. Thus, water suppliers have little choice but to adopt curative strategies and technological fixes for the problem. Curative strategies are already widely used in England and will be needed on a much larger scale in the future if nitrate concentrations in water sources continue to rise. In 1990, curative measures were used to reduce nitrate concentrations in 17 supplies serving 1.7 million people in England (Drinking Water Inspectorate, 1992).

Some aspects of the institutional arrangements for nitrate pollution are flexible and policies and programmes can be modified. For example, experimental measures were introduced through the pilot NSAs scheme because of the considerable uncertainty regarding the effectiveness of land use and land management controls. The NSA Scheme enabled policy makers to test agricultural controls under real farm conditions before implementing them on a larger scale. In addition, some aspects of the EU Nitrate Directive are flexible. There are strict regulations regarding the designation of NVZs, but Member States have considerable discretion regarding the implementation of controls for agricultural activity. For example, Member States will not be forced to implement measures which might reduce the productivity and profitability of farms. Furthermore, the

Member States are required to review their programmes for nitrate pollution management in NVZs at least every four years. As such, there will be opportunities to strengthen the arrangements for the prevention of nitrate pollution if the existing land use and land management controls prove to be ineffective.

Equity, efficiency and effectiveness

The institutional arrangements for management of the nitrate problem were strengthened after 1985. However, a number of substantive issues were not resolved and this has prevented further progress. Agreement has not been reached among land and water interests regarding the equitable allocation of costs. Reliance upon curative strategies and technological fixes for the nitrate problem has meant that the water industry has incurred substantial costs. Since the privatization of water services in 1989, these costs have been added to consumer charges. For example, average household charges for water and sewerage services increased by 5% above the rate of inflation between 1989 and 1992. The agricultural sector has not incurred significant costs because policies and programmes to prevent nitrate pollution have included payments to farmers to compensate for the value of lost production. While policy makers have been reluctant to apply the polluter pays principle in this case, the water industry has continued to argue that it should not be responsible for costs which are generated by the agricultural sector.

There is also disagreement regarding the efficiency of different strategies for nitrate pollution. For example, agricultural interests have argued that source replacement, blending and ion-exchange denitrification may be more cost-effective answers for the problem than changes to farming practices. While cost-benefit analysis was used to examine the efficiency issue, clear answers were not produced (DoE, 1988; Severn-Trent Water, 1988). Comparison of the costs of alternative strategies proved to be difficult because of variations in agricultural and geologic conditions among river catchments. In addition, different results were produced when potential reductions in subsidies for agriculture were taken into account.

A number of concerns remain regarding the effectiveness of the arrangements for the prevention of nitrate pollution. For example, the lack of control over the disposal of farm waste to land is a significant weakness. There is also concern regarding the likely effects of the EU Nitrate Directive. Specifically, further nitrate pollution in NVZs may occur because the Member States are not required to introduce measures which go beyond good agricultural practice. In the UK, good agricultural practice is defined solely on the basis of agronomic and economic criteria and as such may not be consistent with 'good environmental practice'.

Conclusions and recommendations

After 1985, there were a number of significant changes to the institutional arrangements for nitrate pollution in England. Key developments included the introduction of stringent standards for the quality of drinking water, the formation of an inter-departmental group to examine the problem, the establishment of NSAs, and the publication of a Code of Good Agricultural Practice specifically for the protection of water. However, it is evident from this evaluation that there are a number of weaknesses which need to be addressed.

Coordination and cooperation among agricultural and water management officials was particularly limited at the regional level. It is suggested that the situation could be improved by the development of processes and mechanisms to link decision making in the two sectors. For example, a system of policy referral involving the DoE, MAFF, EA, local and regional government, farming organizations, water companies, and other interests could be used to build consensus regarding proposals for land and water management. In addition, coordination could be improved by extending the process used by the EA to produce LEAPs to include land use and management issues. LEAPs could form the basis for a 'shared vision' for land and water if representatives for agriculture and other land-use interests were included in the planning teams.

Integration could be enhanced by providing further opportunities for participation in decision making. This could be achieved using existing consultation mechanisms. For example, the membership of the MAFF Regional Panels should include representatives from corresponding Regional Rivers Advisory Committees (RRACs). Similarly, RRACs should include representatives from the Regional Panels.

The balance between the use of curative and preventative strategies for nitrate pollution was improved after 1985. However, it is clear that additional land-use and land management controls will be needed if further nitrate pollution is to be prevented. While expansion of the NSAs scheme was a positive development for the protection of water sources, water resources should also be protected to ensure adequate supplies for future use. The NVZs are unlikely to provide adequate protection for water resources unless measures beyond good agricultural practice are incorporated into the programme. This limitation could be addressed by targeting agricultural extensification policies, such as set-a-side, at areas which are susceptible to nitrate pollution. Thus, both the production of surplus agricultural commodities and the risk of pollution would be reduced. In addition, farm waste management plans could be developed to ensure that the size of livestock herds is consistent with the area and type of land which is available for waste disposal. It is suggested that financial assistance towards the cost of producing such plans should be provided by MAFF, and the EA should be given opportunities to review them before they are approved.

An explicit and deliberately adaptive approach should be developed because there will always be uncertainties regarding the causes, effects, consequences and management of the nitrate problem. Understanding of the problem and its management could be improved through greater use of experimental control measures. The key is to ensure that the results from the experiments are fed back to decision makers so that adjustments can be continually made to policies and programmes.

The evaluation also indicated that substantive issues regarding equity, efficiency and effectiveness have not been resolved. However, progress could be made in this area by encouraging decision makers to use bargaining and negotiation to address issues such as the allocation of costs, and development of an efficient and effective mix of strategies. Fisher and Ury (1983) developed one approach for effective negotiation which is based on four principles: separate the people from the problem; focus on interests not positions; invent options for mutual gain; and insist on objective criteria. It is argued here that an equitable, efficient and effective solution for the nitrate problem could be reached if these principles were adopted by decision makers.

Acknowledgements

This chapter is a modified version of Watson, N., Mitchell, B. and Mulamoottil, G. (1996), 'Integrated Resource Management: Institutional Arrangements Regarding Nitrate Pollution in England', *Journal of Environmental Planning and Management*, Vol. 39, No.1, pp.45-64. The research was completed with financial support from the Social Sciences and Humanities Research Council (SSHRC) of Canada.

References

Ackoff, R.R. (1974), *Redesigning the Future*, Wiley: New York.

Addiscott, T. (1988), 'Farmers, Fertilizers and the Nitrate Flood', *New Scientist*, October 8, pp.50-4.

Addiscott, T. and Powlson, D. (1989), 'Laying the Ground Rules for Nitrate', *New Scientist*, April 29, pp. 28-9.

Born, S.M., and Rumery, C. (1989), 'Institutional Aspects of Lake Management', *Environmental Management*, Vol. 13, No. 1, pp. 1-13.

Born, S.M., and Sonzogni, W.C. (1995), 'Integrated Environmental Management: Strengthening the Conceptualization', *Environmental Management*, Vol. 19, No. 2, pp. 167-81.

Comly, H.H. (1945), 'Cyanosis in Infants Caused by Nitrates in Well Water',

Journal of the American Medical Association, Vol. 129, pp. 112-16.

Department of the Environment (1986), *Nitrate in Water*, Report of the Nitrate Coordination Group, HMSO: London.

Department of the Environment (1988), *The Nitrate Issue: A Study of the Economic and other Consequences of Various Local Options for Limiting Nitrate Concentrations in Drinking Water*, HMSO: London.

Deyle, R.E. (1995), 'Integrated Water Management: Contending with Garbage Can Decision making in Organized Anarchies', *Water Resources Bulletin*, Vol. 31, No. 3, pp. 87-98.

Dorcey, A.H.J. (1986), *Bargaining in the Governance of Pacific Coastal Resources*, University of British Columbia: Vancouver.

Downs, P.W., Gregory, K.J. and Brookes, A. (1991), 'How Integrated is River Basin Management?', *Environmental Management*, Vol. 15, No. 3, pp. 290-309.

Drinking Water Inspectorate (1992), *Nitrate, Pesticides and Lead 1989 and 1990*, Drinking Water Inspectorate: London.

Fisher, R. and Ury, W. (1983), *Getting to Yes; Nogotiating Agreement Without Giving In*, Penguin Books: London and New York.

Gunderson, L.H., Holling, C.S. and Light, S.S. (1995), *Barriers and Bridges to Renewal of Ecosystems and Institutions*, Columbia Press: New York.

Holling, C.S. (1978), *Adaptive Environmental Assessment and Management*, John Wiley: Chichester.

House of Lords (1989), *Nitrate in Water, 16th Report of the Select Committee on the European Communities, 1988-1989 Session*, HMSO: London.

Ingram, H.M., Mann, D.E., Weatherford, G.D. and Cortner, H.J. (1984), 'Guidelines for Improved Institutional Analysis in Water Resources Planning', *Water Resources Research*, Vol. 20, No. 3, pp. 323-34.

Lee, K.N. (1993), *Compass and Gyroscope; Integrating Science and Politics for the Environment*, Island Press: Washington and Covelo.

Lundquist, J., Lohm, U. and Falkenmark, M. (eds) (1985), *Strategies for River Basin Management- Environmental Integration of Land and Water*, D. Reidel: Dordrecht, Boston and Lancaster.

Magee, P.N. (1982), 'Nitrogen as a Potential Health Hazard', *Philosophical Transactions of the Royal Society of London*, Vol. B296, pp. 543-50.

Mann, D.E. (1983), 'Research on Political Institutions and their Response to the Problem of Increasing CO_2 in the Atmosphere', in Chen, R.S., Boulding, E. and Schneider, S.H. (eds.), *Social Science Research and Climate Change: An Interdisciplinary Appraisal*, D. Reidel: Dordrecht.

Margerum, R.D. and Born, S.M. (1995), 'Integrated Environmental Management: Moving from Theory to Practice', *Journal of Environmental Planning and Management*, Vol. 38, No.3, pp.371-91.

Mitchell, B. (1986), 'The Evolution of Integrated Resource Management', in

Lang, R. (ed.), *Integrated Approaches to Resource Planning and Management*, University of Calgary Press: Calgary, Canada.

Mitchell, B. (Ed.) (1990), *Integrated Water Management: International Experiences and Perspectives*, Belhaven: London and New York.

Mulford, C.L. and Rogers, D.L. (1982), 'Definitions and Models', in Rogers, D.L. and Whetten, D.A. (Eds.), *Interorganizational Coordination; Theory, Research and Implmentation*, Iowa State University Press: Ames, USA.

National Rivers Authority (1993), *Catchment Management Planning Guidelines*, NRA: Bristol.

Newson, M.D. (1992), 'Land and Water: Convergence, Divergence and Progress in UK Policy', *Land Use Policy*, Vol. 9, No. 2, pp. 111-21.

OECD (1989), *Water Resource Management: Integrated Policies*, OECD: Paris.

Severn Trent Water (1988), *The Hatton Catchment Nitrate Study*, Report of the Joint Investigation on the Control of Nitrate in Water Supplies, Severn Trent Water: Birmingham.

Slater, S., Marvin, S. and Newson, M. (1994), 'Land Use Planning and the Water Sector', *Town Planning Review*, Vol. 65, No. 4, pp. 375-97.

Trist, E. (1983), 'Referent Organizations and the Development of Inter-organizational Domains', *Human Relations*, Vol. 36, No.3, pp.269-84.

Walters, A.H. (1984), 'Nitrate and Cancer - A broader View', *The Ecologist*, Vol. 14, No.1, pp.32-7.

Walters, C. (1986), *Adaptive Management of Renewable Resources*, MacMillan: London and New York.

Walters, C. and Holling, C.S. (1990), 'Large-scale Management Experiments and Learning By Doing', *Ecology*, Vol.71, No.6, pp.2060-8.

Watson, N. (1992), 'Coping with the Uncertainties and Conflicts of Nitrate Pollution in the UK', in Shrubsole, D. (ed.), *Resolving Conflicts and Uncertainty in Water Management*, Canadian Water Resources Association: Cambridge, Ontario.

Watson, N., Mitchell, B. and Mulamoottil, G. (1996), 'Integrated Resource Management: Institutional Arrangements Regarding Nitrate Pollution in England', *Journal of Environmental Planning and Management*, Vol. 39, No.1, pp.45-64.

7 Coherence and divergence in Dutch physical planning and water management planning

Marius Schwartz

Introduction

The perception of the meaning of 'planning' and, in particular, the way in which the process of planning should be shaped, is to a large extent determined by social and cultural factors. This is the reason why members of different societies have developed different ideas about the most appropriate ways of planning. Of course, theories about planning need to be applicable to each society, but these theories will be adapted according to social and cultural circumstances.

In the Netherlands social, cultural and physical circumstances have led to strong governmental influence over planning, especially when compared with other Western countries. This influence over planning is reinforced by extensive legislation in the field of planning which, in turn, further facilitates governmental influence.

Governmental influence is, of course, not the only reason for the existence of extensive legislation; indeed, the main reason is the general need for well-balanced and transparent planning on a number of important topics. Whilst some of these topics are dealt with in separate fields of planning, several topics demonstrate that coherent planning is necessary. Although the existence of different planning systems might facilitate this coherence, such a division of functions can also lead to divergence.

The need for coherent planning is illustrated in the following section of this chapter; in this section the need for the coordination of plans is discussed. In the third section, three related fields of planning (these are, physical, water management and environmental planning) are outlined. Examining the legislation on the three planning systems and the potential for 'interference' between the fields of planning, it appears that the existence of three planning systems possibly could offer the opportunity for coherent planning. However, does coherent planning occur in practice, or does the presence of three planning systems lead to divergence? These questions are examined in case studies of two zoning plans and two water management plans and though comparison conclude that in practice divergence in planning occurs. Explanatory factors for this divergence are presented. The chapter ends with a consideration of some current developments that may lead to more coherent planning for the Dutch environment.

Water management and physical planning: related topics and problems

In the Netherlands policy on water management has been formulated and expressed on a separate planning system. The justification for the existence of this separate system is the great influence of water management over land use: when water management fails, safe living and development is impossible. Although this is the fundamental explanation for the existence of a separate planning system, from an institutional point of view it can be argued that the established civil organisation of rijkswaterstaat and the local water boards has created a powerful lobby in favour of a separate water policy and planning system.

However, seen from a water management point of view, several topics and aspects of water management are strongly related to other policy fields, especially those of physical planning and environmental planning. Conversely, physical planning needs effective water management for the realisation of its targets.

Seen from a physical planner's point of view, the following policy and practice developments militate in favour of cooperation with water management planners:

- in lowland rural areas, the (spatial) development of different types of land use depends highly on water management. Water levels and water quality influence the functioning of these types of land use to a large extent;
- the urban water front development, which started in the Netherlands in the late 1980s, caused urban planners to pay increased attention to water. Old harbours and backstreet districts were transformed in order to create up-market housing and office areas (as in Amsterdam, Rotterdam, Roermond, 'Drecht'-cities). In these circumstances urban planners expect water managers to provide and maintain clean surface water, preferably with attractive banks for recreational purposes;
- in search of new planning concepts that consider landscape values, physical planners adopted surface water and ground water systems as bases for spatial design (see for example: Ministry of Housing, Physical Planning and the Environment, 1989);
- the 1989 Law on water management obliged the coordination of water management and physical plans. In part, this coordination is determined by law, but this only applies to the content of both systems.

86

Seen from a water management planner's point of view, the following developments require cooperation with physical planners:

- if water policy has to realize a certain function for a water system, contribution from physical planners may be necessary. Protection of a sensitive ecosystem, such as a spring, requires upstream spatial measures that protect the ground water system from polluting the spring with nutrients from the use of manure and fertilizers in agriculture;
- water managers, by themselves, cannot attain many of their water quality objectives because their own legal instruments are insufficient. The reason for this is that a significant element of water pollution is caused by diffuse sources, such as traffic, agriculture, the use of household products and leaching of materials (Ministry of Transport and Public Works, 1994). The reduction of emissions from these sources also requires action by physical and environmental planners;
- in the mid-1980s new policy concepts were introduced: the water system approach and comprehensive water management. Both concepts aim to provide a water management system based on a more comprehensive view of the environment. Part of this can be achieved through closer cooperation with municipal authorities, paying attention to the relationship between water bodies and their environment;
- the introduction of the above-mentioned Law on the water management implies coordination with other planning systems;
- some specific water management problems such as sludge deposits and sewage treatment plants, suffer from the NIMBY syndrome and therefore require cooperation with physical planners.

The above-mentioned topics point to the conclusion that a coherent policy on water management, environmental and physical planning is necessary. All of the examples demonstrate the interrelationships between the various planning fields. Poor coordination of plans will lead to policy targets that cannot be achieved, projects that are unlikely to be realized and will lead to frustrated policies and employees.

However, to what extent is coordination facilitated by the operation of the planning systems? The legal aspects of this topic will be dealt with in the next section of this chapter.

Physical planning, environmental planning and water management planning in the Netherlands

The legal background of the most important Dutch plans for the environment is laid down in three planning laws:

- the physical planning system (which has its legal foundation in the Law and Regulations on Physical Planning);
- the environmental planning system (which has its legal foundation in the Law on Environmental management);
- the water management planning system (which has its legal foundation in the Law on water management).

The key points of these planning systems are outlined in Table 7.1 and the relationship between these systems is shown in Figure 7.1. From this table and the figure some similarities can be identified between the planning systems:

- all plans have an impact on the physical environment;
- all plans aim to ensure actions in the public domain and because plans are the product of a democratic controlled organisation, the range of interests involved is considered and balanced in a plan;
- the three planning systems all prepare plans at each governmental level.

These similarities stress the need for coordination of these plans. When the regulations were drafted, the need for coordination was recognised. On several points this has led to the inclusion of coordination measures in the regulations. Figure 7.1 illustrates a scheme in which the legal relationships within and between the planning systems are outlined.

A further focus of the physical and water management planning system indicates that differences occur concerning the aim of planning (structuring spatial development versus establishing functions of water systems) and the legal status of plans. In addition, differences occur in the training and perceptions of planners; this can be seen in the contrast between physical planners versus water management planners who have a background in civil engineering.

Now that the key features of the Dutch planning systems have been explained a more detailed analysis can be presented of water management and physical plans at local level. Because both plans are the subject of a case-study later, the most relevant characteristics are given first.

The main characteristics of a water management plan include:

- it provides a framework for the activities of the (local) water board; the plan indicates what activities a water board is undertaking in order to fulfil its task;

Table 7.1

Main features of the physical, environmental and water management planning systems in the Netherlands

Planning System	Objectives of the Planning System	Jurisdiction	Main features plan	Revision period years[1]
Physical planning	Quality of the physical environment and spatial development	National	*Headlines spatial policy: strategy on spatial development in urban and rural areas *Key Planning Decision: binding for the National government	set in plan
		Province	*regional allocation of land use *strategy on spatial development *tuning with environmental and water management plan	10
		Municipality[2]	*detailed allocation of land use *the zoning plan is binding by law for citizens, firms, etc.	10
Environmental Planning	Protection of the environment	National	*headlines environmental policy *strategy to reach targets	4+2
		Province	*obligatory annual operational programme *headlines on provincial policy *allocation of groundwater protection areas, silence areas and soil protection areas	4+2
		Municipality	*obligatory annual operational programme *tuning with provincial physical and water management plan *facultative strategic plan	
Water Management Planning	Co-ordinated and functional water management	National	*obligatory annual operational programme *global allocation of functions of state waters *national qualitative and quantitative targets(inclusive floods)	set in plan
		Province	*strategy on those targets *allocation of functions to water systems *strategy on those targets *tuning with provincial physical and environmental plan	4+4
		Waterboard	*(detailed) allocation of functions *operational framework for activities waterboard	4+4[3]

1 The period as laid down by law; the number of years appearing after a '+' is a legal option to lengthen the revision period.
2 In fact two types of plan exist for the jurisdiction of the municipality: the zoning plan and the regional plan. The latter is less specific and detailed as the zoning plan, it has more strategic features, a visionary type of spatial plan. The described features in the scheme refer to the zoning plan.
3 In fact the revision period of the water management plan is set in a provincial directive; most directives use the '4+4' period.

Figure 7.1   The relationship between planning systems in the Netherlands

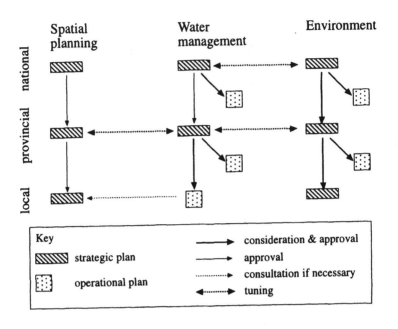

- the plan has to consider the provincial water management plan, which has more strategic features and allocates functions to waters, for example, a river may have a primary function for shipping, or a complex of ditches in a polder may have an agricultural function; these allocated functions are elaborated in a water board plan.

The main characteristics of a zoning plan include:

- a municipality uses the plan as the basis for the assessment of building licences (such licenses operate under the jurisdiction of the Law on Physical Planning);
- the allocation of land use; the law does not make any explicit differences between allocations for land use or water: 'ground includes water';
- it is the only plan that is legally binding on citizens and companies;
- during the preparation of a zoning plan the municipal authorities *may* consult the water board, if necessary.

From these characteristics the following conclusions can be drawn:

- for the same location both types of plan consider the range of interests involved and try to balance them resulting in the allocation of land uses (physical planning) or functions (water management);
- only the zoning plan is binding on citizens; citizens can appeal against a zoning plan, not against a water management plan;
- the water management plan is binding on the water board; for this reason the water board is obliged to implement the plan in an active manner, but the municipality is not obliged to implement the zoning plan in an active manner, passive enforcement is sufficient.

The water management plan of the water board and the zoning plan of the municipality offer the opportunity for a coherent policy to be established for both policy fields. From this it can be seen that in the same location a variety of different aspects of policy that are important for the physical environment can be developed via these two separate instruments. An important pre-condition for the establishment of coherent policy is that the contents of these plans should be coordinated. This attempt to provide coherence will be examined in the case study.

Case study

The contents of a water management plan and a physical plan were compared for two territories in the Netherlands; both of these territories are in the eastern part of the Netherlands (Schwartz et al., 1995). Some characteristics of the case studies are given in Table 7.2.

These two territories were chosen mainly because of the fact that the zoning plans had been recently published (respectively 1992 and 1994), and thus any new ideas regarding concepts of planning and new policy, for example the need to pay attention to the relationship with other planning systems, should have been incorporated in these recently published plans.

Comparing these plans a number of practical aspects were identified which hamper any comparison:

- the plan territories vary: the demarcation of the zoning plan is based on the administrative borders, whilst that for the water management plan is based on water system borders;
- the scale of the maps vary;
- the background of the maps vary.

Table 7.2

Characteristics of the two case studies

Location	Plan title	Jurisdiction	Published
Warnsveld North eastern part of province of Gelderland	Zoning plan	Municipality	1992
	Comprehensive water management plan for Eastern Gelderland	Water board	1994
	Regional plan for Eastern Gelderland	Province	1987
	Water management plan	Province	1991
Denekamp Eastern part of the province of Overijssel	Zoning plan	Municipality	1994
	Water management plan	Water board	1989
	Regional plan for Twente	Province	1990
	Water management plan for Overijssel	Province	1991

Regarding the content, two more fundamental problems occurred which might interfere with the coordination of plans. One problem is that for the same territory different allocations occur in the plans: the zoning plan allocates agricultural land use, while, at the same location, the water management plan allocates a nature protection function. Although this example is extreme, more subtle examples can be found elsewhere. Table 7.3 demonstrates this type of allocation problem.

More subtle examples of mismatches are illustrated in Table 7.4. Although apparently the terminology used to determine allocations is the same in both plans, from an examination of the definition of aims it is possible to detect subtle differences. If these plans are implemented on the basis given then a number of interpretation problems may arise. For example, a construction license or permission for an agricultural activity may be given based on the zoning plan, while at the same location the water board may wish to raise the ground water level in order to ensure the conservation of specific natural features.

92

Table 7.3

Differences in allocation of land use and function between physical plans and water management plans

Type of plan:	LOCAL			PROVINCIAL		
	Zoning plan	Water management plan	Regional plan		Water management plan	
Description of land use or function:	Agricultural use	Water for agricultural use	Interrelated agricultural use and protection of nature values		Water for agricultural use and nature dependent on percolation water	
Aim:	Continuation of agricultural land use. New farms can be started (by dispensation). Protection of existing landscape values via construction permits.	Protection of basic water quality targets, minimize the surplus and shortage of water, surface water primarilyy for maintenance of ground water levels, development of corridors, isolation of special nature elements.	The policy for these areas aims at continuation and development of agricultural use taking into account the interrelated nature and landscape values.		Water levels that aim at protection and restoration of special nature values and, as far as a combination is possible: maintaining ground water level, sprinkling etc.	
Comparing the aims:	From the water management plan it appears that the function 'agricultural use of water' also includes two sub aims on nature development, which not at all are covered by the country plan.		From the aims in both plans a clear difference in priority appears: the water management plan prioritizes a nature orientated water level, while the regional plan prioritizes the continuation and development of agriculture.			

Table 7.4

Differences in overall aims between physical plans and water management plans

Type of plan:	LOCAL		PROVINCIAL	
	Zoning plan	Water management plan	Regional plan	Water management plan
Description of land use or function:	Agricultural area	Water for agriculture and nature (not dependent on percolation water)	Interrelated agricultural use and protection of nature values	Accounting water quantity because of downstream nature
Aim:	Continuation of agricultural land use. New farms can be started (by dispensation). Protection of existing landscape values via construction permits.	Protection of basic water quality standards and restoration of special nature values by isolation and/or maintenance of high water levels, as far as possible in combination with agricultural use. Development of nature friendly banks (corridors).	The policy for these areas aims at continuation and development of agricultural use taking into account the interrelated nature and landscape values.	Water levels that protect sensitive downstream nature, of quantity and co-ordination of location of ground water extractions regarding the protection of downstream nature.
Comparing the aims:	No co-ordination, potential conflict: The country plan primarily aims at continuation of agricultural use, while the water management plan, although not excluding this function, primarily focuses on the development and restoration of nature values.		No co-ordination, potential conflict: Although the regional plan takes into account the interrelations with nature values, the water levels that aim at protection of downstream nature as pointed out in the water management plan, will interfere with the agricultural land use.	

Explanatory factors

Although the existence of different types of plan offers the opportunity to provide a coherent policy for the environment of rural areas, on which both the municipality and the water board can base their (licensing) activities, divergence can result if poor coordination of the plans occurs. This was illustrated in the case study.

A number of different reasons can be detected which explain these difficulties (Glasbergen et al., 1988; Brussaard et al., 1995). They include differences in law and procedures, organisation and management, and the objectives of planning.

Law and procedures

The procedures utilised by the physical and water management planning system are not in step with regard to the periods used for revision (see Table 7.1). Policies change over time resulting in different degrees of achievement of the plans. The laws for both planning systems include a procedure by which the most recent provincial plan contains the current policy, but this regulation can lead to a lack of clarity.

The only procedure laid down in *The Law on Physical planning* to ensure coordination at a local level between a water board and a municipality is the above-mentioned requirement to 'consult the water board if necessary'. However, this can generate a noncommittal attitude from both a water board and a municipality.

A third aspect is the different nature of the plans; as has been explained above, the zoning plan and water management plan are used for different purposes.

Organisation and management

Traditionally water management and physical planning have been undertaken by separate departments. Given the different background of planners (see next section) this leads to a gap between the two fields of planning. For the water boards and municipalities this is also an institutional problem because these bodies are separate organisations.

Objectives of planning

Surface water and ground water are dynamic features - they flow - whereas land and soil are static. Although this distinction is obvious, it does lead to different approaches to the task of planning. This is associated with another source of mismatches, and this one has its origin in the background and training of planners. Generally, those involved in water management have a technical education, whilst those in physical planning have a social science education. These backgrounds lead to different approaches to planning methodology and procedures.

Towards better coordination

Although the existence of three planning systems makes coherent planning for the environment complicated, on the other hand it also offers several opportunities for the realisation of policy: three laws provide more instruments then one. However, in practice, divergences in the theory and method of planning can generate a lack of clarity in the development and implementation of policy (see Table 7.3 and 7.4).

Several reasons have been provided which offer explanations for this divergence, but the extent and influence of these separate reasons is unclear. As a consequence the measures adopted in order to improve policy coordination consider all possible causes. More fundamental research is needed in order to identify which reasons are the most important and which factors are of less significance.

A number of efforts have been made to attempt to enhance the coordination of plans. A selection of developments at provincial and local level are presented below.

Law and procedure:

Provincial plans The contents of the various provincial plans could be better coordinated if the procedures for plan production were synchronised. Several provinces, for example, Groningen and Gelderland, have prepared and published their regional, water management and environmental plan alongside each other. All three plans contain the most up-to-date policy and share a number of procedural characteristics.

However, planners realize that the synchronisation of plan preparation is not sufficient, the next step is the integration of such plans; this objective has been stated by several provinces, for example, North-Holland, Groningen, Drenthe and Gelderland. It is expected that all provinces will follow this example when the first results of the experiment undertaken by the progressive provinces become clear.

How this integration will be achieved is not yet clear. Most probably an overarching strategic plan or vision will be developed which will form the basis for the development of the three subject plans. Because these plans will have the same starting point, it is anticipated that the coordination will be achieved. An important aspect which requires further research is the legal status of the plans and the prescribed procedures, which are not the same for all plans.

Water management plan and zoning plan Because there are institutional differences between the water boards and the municipalities, the integration of plans is inhibited. Despite the fact that changes will be made to the laws in order to enhance interpretation, the question remains: will a change in the law enhance coordination when the law does not specify which aspects of plan preparation and complementation should be coordinated?

Organisation and management:

Provinces The water management and physical planning took place in separate departments of the provincial authorities. The trend is integration of these departments into a 'planning department' in which water management, physical planning and environmental planing are integrated.

Water boards and Municipalities Institutional differences, partly laid down in law, make the merger of these two separate organisations very unlikely. The different nature of the organisations has always created problems in ensuring cooperation and coordination.

An important organisational change that has recently occurred is the merging of water quantity boards and water quality boards. These two types of water board were separate organisations. The quantity boards, in particular, often had jurisdiction over a very small territory. As a result of a policy that is aimed at the encouragement of comprehensive water management, these two types of water board have been merged. The single integrated water boards have become large organisations with a range of planning experts that now have the knowledge necessary in order to coordinate water plans with spatial plans. Small water boards were not equipped for those purposes.

Objectives of planning

One way to obtain better coordination is to introduce and encourage the development of new planning concepts. A concept that can be used in both water management and land use planning is the water system approach. By using the same design concept for both types of plans, the objectives of planning (land use and water) are approached from the same direction and in the same way.

Another 'new' concept is the area oriented approach (Bouwer, 1994). In this concept the objective of planning is not land use or water, but the range of problems encountered in a specific territory. The starting points of such a planning methodology is a consideration of the interests represented in the area: are the interests of actors and agencies threatened, do they need to be protected, do they have specific aspirations and can such aspirations be satisfied? If consensus on these aspects can be reached, then a comprehensive plan can be prepared in which the achievement of aspirations can be specified. A final step is the preparation of the water management and land use plans according to the principles and objectives of the comprehensive plan.

Conclusion

In the Netherlands, the physical planning system, environmental planning system, and water management planning systems demonstrate coherence in the impact they have on the environment. This is evident at national, provincial and local levels. However, the relationship between the topics dealt with by these three fields of planning also necessitates the development of coherent policy through the coordination and tuning of plans; this requirement is, in part, laid down in the regulations for these planning systems.

In order to investigate the coherence of planning in practice, two case studies have been carried out. From these case studies the conclusion can be drawn that significant divergences occur with respect to both the allocation of land uses and the definition of the aims expressed by separate plans prepared for the same territories.

This divergence between the various forms of planning is not unique to these two cases, rather it represents a major issue for Dutch environmental planning. In practice a number of different developments can be observed which are aimed at the generation of a coherent form of planning for the environment.

A number of trends can be identified in the measures which have been adopted. Several provinces have merged their planning departments, whilst at the same time, different methods for the integration of plans have been adopted - in these experiments the area oriented approach is an important concept.

At the level of the municipality and the water board, cooperation in the preparation and implementation of plans and projects (area oriented, comprehensive water management) tends to be the preferred solution in order to achieve a more coherent policy outcome.

The variety of measures that have been taken indicate the growing awareness among planners and the other actors and agencies who are involved, that things have to change. This awareness is the starting point for a more coherent approach to planning for the environment.

References

Bouwer, K. (1994), 'The integration of regional environmental planning and physical planning in the Netherlands', *Journal of Environmental Planning and Management*, Vol.37, No.1, pp.107-16.

Brussaard, W., van der Velde, M., Blom, C.J., Haerkens, H.M.J., Hagelaar, J.L.F., Ovaa, B.P.S.A, van der Vlist, M.J. (1995), *Een brede kijk op waterbeheer, een juridisch-bestuurlijke evaluatie van het instrumentarium van de Wet op de waterhuishouding* (A comprehensive view on water management, a juridical-administrative evaluation of the instruments of the Law on the water management), Landbouwuniversiteit Wageningen - Project team NW4, Wageningen.

Glasbergen, P., Wessel, J., Baltissen, J.H.P., Compaijen, C.J., Groeneberg, M.C., Kuijpers, C.B.F. (1988), *Samenhang en samenspel in het waterbeheer, het streven naar integraal waterbeheer* (Coherence and team-work in water management, the struggle towards comprehensive water management), Delftse Universitaire Pers, Delft.

Ministry of Housing, Physical Planning and the Environment (1989), *Fourth report (extra) on physical planning in the Netherlands* (English summary), Lower House of Parliament 1990-1991,21879, NR.5-6.

Ministry of Transport and Public Works (1994), *Evaluatientota Water* (Evaluation Report Water), Lower House of Parliament 1993-1994, 21250, NR.27-28.

Schwartz, M.J.C., Leijen, I., Hengeveld H. (1995), 'Afstemming bestemmingsplan en waterbeheersplan is vrijblijvend' (Tuning of zoning plan and water management plan is noncommittal), *Stedebouw en Volkshuisvesting*, Vol.76, No.1/2, pp.28-33.

8 Risk, environment and land use planning: an evaluation of policy and practice in the UK

Gordon Walker, Derek Pratts and Mark Barlow

Introduction

An important component of the current environmental agenda revolves around questions of risk. A range of problematic environmental and societal concerns have been characterized as having major risk dilemmas at their core (Beck, 1992; 1995). Environmental policy in the UK has consequently taken on board explicit approaches to risk assessment in an attempt better to manage and control perceived threats to health and environmental quality (Department of the Environment (DOE), 1995; Interdepartmental Liaison Group on Risk Assessment, 1995).

Land use planning is one of a number of areas of public policy to have a role in the management of environmental risks. Planning is relevant to risk management because environmental risks typically have spatial consequences. In terms of simplified models of the risk management process (O'Riordan, 1979), the land use planning role may be characterized as one of risk 'evaluation' and risk 'control'. The first two stages of risk management (identification and estimation) provide 'scientifically' derived information for the more political steps of deciding the acceptability of risks (evaluation) and the appropriateness of action to control risks. Such decisions can be highly problematical having to cope with often uncertain science, conflict between 'expert' scientific and lay public perceptions of risk, and difficult balances between risk and other planning concerns (Walker, 1991; Owens, 1994).

The aim of this chapter is to evaluate the role for land use planning in contributing to the effective management and control of environmental risks. Whilst the full range of risks with which the land use planning system is concerned would extend to include 'natural hazards' (for example flooding) and risks associated with traffic safety, this discussion is limited to risks of a technological origin associated with some form of potential or actual environmental pollution. Our evaluation draws, in particular, on research into land use planning control of two forms of risk - industrial major accident hazards and contaminated land. These are both widespread sources of risk with significant land use implications. The structure of the chapter is as follows. First, a broad review of central government policy guidance is provided, identifying established, recent and proposed planning roles, and the types of environmental risk addressed and methods of intervention available in each case. Second, industrial major accident hazards and contaminated land risks are examined more closely, in each case reviewing the development of relevant planning policy. Third, a number of complexities and constraints are identified which have limited the effectiveness of planning action. Finally, conclusions are drawn regarding the potential for planning control of risk, and the position of planning within the modernized structures of environmental risk management emerging in the late 1990s.

Environmental risks and planning roles

Since the early 1970s land use planning has gradually taken on board a range of pollution and environmental risk concerns (Wood, 1989; Owens, 1994; DOE, 1992a), in recognition of its role in determining spatial relationships between sources of risk and potential victims, its inherently preventative power to stop or modify new development and its degree of openness to public participation compared to other areas of pollution and risk control.

The current policy position across a range of environmental risks can usefully be reviewed by utilising Fischoff et al.'s (1978) approach to categorising points of intervention along a risk pathway, extending from the source of risk to the eventual target under threat of harm or damage. As shown in Table 8.1, this analysis reveals four different points at which land use planning powers can potentially be used to control environmental risks. Also indicated in Table 8.1, for ten different sources of risk, is the extent to which these points of intervention have been recognized in land use planning policy and whether these are established or more recent policy developments.

The ability to regulate the spatial location of sources of risk has been and continues to be the main anticipatory instrument (Table 8.1, column 1). The area of waste disposal provides an example where this established role in regulating risk through controlling the location of risk sources has been strengthened by emerging legislation and policy guidance. In the case of waste disposal by incineration, whilst the impacts and the potential risks deriving from emission will continue to be within the remit of the pollution control authorities (PCAs), the local planning authorities (LPAs) will be able to take account of the cumulative

Table 8.1
Land use planning and risk concerns

Planning Considerations

Risk Concerns	Location of new sources of risk	Risk reduction at source	Location of population at risk	Provision of protection for population at risk
Local / regional air pollution (fixed sources)	E	e	R	e
Local / regional air pollution (mobile sources)	E/R	R	E/R	-
Global atmospheric pollution	-	R	-	-
Major accident hazards	E	e	E	e
Contaminated land	-	e/R	e/R	e/R
Waste disposal : landfill	E/R	E	E/R	-
Waste disposal : incineration	E/R	e	E/R	-
Waste pollution (point sources)	E	e	-	-
Waste pollution (non-point sources)	E	e	-	-
Nuclear establishments	E	-	E	-

Key E/e Established concerns (before 1985) Capital letter indicates major role,
R/r Recent concerns lower case letter indicates minor role

impact of emissions, where incinerators are built in close proximity to other sources of pollution. Similarly in the disposal of waste by landfill, whilst the Environment Agency has taken over the powers of the waste regulation and waste disposal authorities (WRAs), the non-metropolitan LPAs have now acquired the additional strategic role of preparing a waste local plan, identifying current and future site of waste disposal authorities (DOE, 1994b; DOE, 1992a).

Risk reduction at source (Table 8.1 column 2) is indicated as less important for most sources of risk, principally because this is where PCAs have the dominant role. As discussed later, planning guidance has emphasized that LPAs should not become involved in regulating sources of risks through the use of technical or technological tools. The scope for intervention is greater when new installations are going through environmental impact assessment (EIA) procedures as negotiation may centre around aspects of risk reduction. The most fundamental intervention in terms of risk reduction is seen in moves to develop an integrated approach between air quality management, transport planning and development control. One of the key objectives is to reduce the need to travel, and hence reduce emissions (Rydin, 1995; DOE, 1995). This creates a risk reduction policy, where the benefits are not just translated to the local environment, but are part of a wider strategy to contribute to national emission reductions as part of international commitments to tackling European and global atmospheric pollution (DOE, 1994c).

Applying controls over the location of populations at risk is a further intervention point for the planning regime (Table 8.1, column 3). The situation generally relates to controlling the incremental development of residential zones in proximity to sources of risk. In the case of nuclear establishments and sites with the potential for major accident hazards, the need to have regard to the risks arising to the general population within 'hazard zones' has been an established feature of the legislation. In the case of waste disposal there has been a statutory requirement for LPAs to consult the then WRAs on development proposals within 250 metres of former and existing landfill sites. However, only recently has such a policy been applied to polluting facilities regulated under the Integrated Pollution Control (IPC) regime, with guidance suggesting that any proposed development within 500 metres of such a facility should be subject to special attention by the LPAs and consultation with the PCA (Her Majesty's Inspectorate of Pollution, 1995). The least developed intervention category relates to actions to provide protection for the population or environment at risk, which are implemented away from the source of risk (Table 8.1, column 4). This minor role in part reflects the limited range of measures which can be taken to protect often diffuse populations and environments and the focus of policy on prevention and 'polluter pays' approaches. However, limited action *can* be taken in some cases, particularly where risk control at source is itself problematical. This typically involves the use of planning conditions to require additions or changes to building design and structure which provide a degree of risk protection for members of the public occupying these buildings.

Case studies of policy development

Two examples of environmental risk concerns which provide both contrasting and common experiences in dealing with risk within the land use planning system, are those of industrial major accident hazards and contaminated land. The next two sub-sections provide a brief review of the background and policy development for each of these risks, forming the basis for later comparative evaluation.

Industrial major accident hazards

Major accident hazards can be broadly defined as the storage or use of toxic, explosive or flammable substances, where in the event of a major accident local people and the nearby environment could be seriously affected. Such accidents rarely happen, but the potential hazard or threat has been demonstrated both in a number of major disasters, for example at Flixborough, Seveso (Italy) and Bhopal (India) (Lagadec, 1982; Walker, 1994a) and in smaller scale incidents at a range of sites in the UK (Health and Safety Executive, 1983; 1993). Sites presenting a potential major accident hazard are broadly spread across the country (see Figure 8.1) although there is distinct clustering in areas which have a concentration of the chemical and/or petrochemical industry (Walker and Draycott, 1995).

The role for planning in the control of major accident hazards is conceptually very simple. Clearly, if despite all on-site safety measures, an accident involving the release of hazardous substances occurs, it will be better if the number of people living or working in the vicinity of the release is small rather than large. As indicated earlier in Table 8.1, achieving some degree of separation between hazard and people, can in principle be achieved through land use planning, through controlling both the location of sources of risk and the proximity of other surrounding development.

Policy recognition of the role for planning in the control of major accident hazards appeared relatively early in 1972, being further highlighted in the wake of the Flixborough accident in 1974. Procedures were established whereby LPAs could consult with the then factory inspectorate (now the Health and Safety Executive or HSE) to obtain expert advice on the safety implications of new hazardous installations, or development in the vicinity of existing ones (DOE, 1972). In practice, at this stage though, planning authorities had inadequate information on existing hazards, limited powers to control new hazards and insufficient guidance as to what constituted 'significant development in the vicinity' (Petts, 1988, p.11; Walker and Macgill, 1985, p.464).

Figure 8.1   Hazardous installations in England

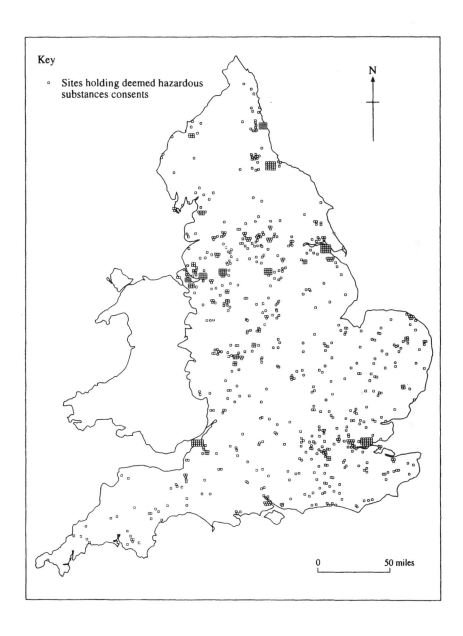

Source: Walker and Draycott, 1995.

Since this time various policy developments have attempted to addressed these deficiencies. There have been two further DOE circulars in 1984 and 1992 whilst the Hazardous Substances Act 1990 and Regulations 1992, attempted to fill various loopholes in control powers, creating hazardous substances authorities (largely LPAs) to which applications for a consent to hold hazardous substances have to be made, in addition to any related application for planning permission (Walker, 1994b; DOE, 1992b).

At an EC level the accident at Seveso in 1976 led, after much deliberation, to the 'Seveso' Directive on major accident hazards in 1982. Despite this being a key piece of legislation for tightening up regulation over, in particular, the higher hazard sites, this Directive did not recognise a role for land use planning. However, the disaster in Bhopal in 1984, which graphically displayed the dangers of poor siting and high population concentrations next to major hazard sites (Walker, 1994a), compelled the European Commission to introduce a planning requirement into a new 'Seveso II' Directive which is expected to come into force in 1998 (Walker, 1991; 1995).

Given the twenty five year period over which the planning role in major hazard control has evolved, it might be expected that problems would now be few and far between. However, whilst planning practice is now on a much firmer footing, recent research has identified a range of ongoing difficulties. For example, a study of the implementation of the Hazardous Substances Act and Regulations undertaken for the DOE (Walker, 1994b) found that there was poor understanding of the regulations and the planning role within LPAs; inconsistencies in the application of planning conditions; inconsistencies in definitions of hazardous sites and exemptions from the regulations; uncertainty and disagreement about the extent of non-compliance; and little active enforcement by LPAs. A survey of LPAs in 1993 also found limited inclusion of policies on hazardous industry in development plans (Walker and Bayliss, 1997) which is particularly significant given recent moves towards more 'plan led' development control. A number of other studies through 1980s and 90s identified complexities in decision-making and constraints on planning action in this area (Walker, 1984; Petts, 1988; Miller and Fricker, 1993).

Contaminated land

Land contamination poses a risk to health and environmental quality through the presence of toxic, flammable or explosive substances. This threat can be in a variety of forms, for example, leaching of chemicals into a watercourse, concentration of toxic substances in plants or the build up of explosive methane gas. A wide range of activities can lead to land contamination, including landfill waste disposal, gas manufacture, metal manufacture and processing and mining, all of which have been carried out in the UK for a long time. It is for this reason that there is a substantial land contamination problem in the UK, although estimates of its extent vary substantially (National Rivers Authority, 1994). As indicated earlier in Table 8.1, planning powers can potentially be used to either keep sensitive developments away from contaminated sites, to require the investigation and treatment of contamination before any development takes place, or to control aspects of building design to limit the ingress of hazardous gases.

In contrast to major accident hazards, policy development for contaminated land has been characterized by controversy and stilted progress over an extended period. The first attempts to deal with contaminated land problems came in the form of guidance notes issued by the DOE in the 1970s. These were intended to provide technical guidance on best practice for dealing with contaminated sites of various sorts - if and when such sites were identified. It was also established at this time that land contamination was a material planning concern. However, a whole series of key questions remained unanswered, including how contaminated land was to be identified and who should pay for remedial treatment. As a result, any action taken by LPAs to deal with contamination at this time was very much on an ad-hoc basis.

Such problems were highlighted in highly critical reports produced by the Royal Commission on Environmental Pollution (RCEP) in 1985 and the House of Commons Select Committee on the Environment in 1990. These prompted the government look more seriously at legislation, so that in the Environmental Protection Act 1990 a provision was included requiring, amongst other things, that LPAs compile public registers of potentially contaminated land in their area. This had the aim of establishing evaluation and action over contaminated land on a firmer information base, so that contaminated sites would be picked up early in planning processes. Details specifying how the compilation of registers would be implemented were put forward in two successive consultation papers in 1991 and 1992. However, a third paper and policy review in 1994 dropped the provision altogether, arguing that it was unsuitable as a way of dealing with the contamination problem in the UK (DOE, 1994a). Instead this paper laid out a range of ways in which it was expected that contamination problems would be addressed, with the due to be formed Environment Agency acting as a centre of expertise and advice, and local authorities playing a key role through both statutory nuisance provisions and planning decision-making. After a further period of deliberation, more detailed regulations were again proposed in 1996 along similar lines to the 1994 paper. Partial guidance for LPAs was provided by advice in PPG23 which reinforced the need for development plans to include policies and detailed criteria for dealing with land contamination problems (DOE, 1992a).

Despite this series of papers and proposals, currently there is still no substantial framework in place for dealing with land contamination in the UK (in contrast to other countries in Europe). Successive policy reversals and delays have meant that LPAs have continued to operate in a very ad-hoc manner (Petts, 1995).

Some LPAs have had active policies trying to identify contaminated land and land requiring remedial treatment, wherever possible supported by grants under derelict land programmes. Others, appear to have essentially ignored the problem with no clear policies contained in development plans and with planning applications being dealt with on a poor information base. Although the latest regulatory proposals may see some improvements in local action, these have been criticized as leaving too much to interpretation at a local level, seeking to deal with only high risk sites and requiring only limited levels of clean-up (Talbot, 1996).

Constraints and complexities for the planning role

In the discussion so far a number of problematical issues have been identified which have featured in attempts to manage risks through the land use planning system. In this section attention is focused on a number of key constraints and complexities, selected for their generic relevance across the range of risks earlier identified. The two case studies introduced in the previous section will be particularly drawn on.

The legacy of historic risks and land use patterns

A fundamental constraint on land use planning is its limited ability to change current patterns of land use. Whilst new development may be controlled, it is rare to find planning powers being used to change existing land uses (where development has taken place) in anything but an incremental manner. This creates a powerful land use inertia which is compounded by the late recognition of most forms of environmental risk, generally long after many of these have become part of established urban and/or rural landscapes. This creates a 'dilemma of control' (Collingridge, 1980, p.16) which can fundamentally frustrate attempts to utilise powers of spatial control to reduce levels of risk exposure.

For industrial major accident hazards there are very few new sources of risk appearing each year. Rather the majority of hazardous sites have been in place for many years. They were established long before planning controls took any specific account of risk concerns and hence substantial populations often accumulated in their vicinity (Walker and Draycott, 1995). Faced with this legacy of poor siting, most LPAs recognise their inability to improve matters significantly - few development plan policies refer to the relocation of hazardous sites as a policy aim (Walker, 1994b). For populations living in the vicinity of hazardous sites, the possibility for action is limited to preventing the accumulation of additional people in areas significantly at risk (i.e. to not let a bad situation get even worse). Whilst it has been HSE and DOE policy to follow this line (HSE, 1989), at local level it can be difficult for planners to justify preventing say a new housing development on safety grounds when there may already be many people living closer to the hazard site and at a greater level of risk. For contaminated land, the historical legacy is fundamental to the problems which have to be faced. The land contamination now found across the UK has been created by pollution and waste disposal often extending back to the 19th century. In many situations contaminated land has been redeveloped and built on in ignorance of the potential risks to health, creating intense contemporary conflicts between the continuation of existing land uses and the taking of any substantial action to reduce or remove the risks involved. It is very rare though to find action being taken, for example, to remove housing or other forms of development because of the contamination of the site they already occupy. Instead, policy is directed at taking action where new development is proposed (through the imposition of planning conditions or the prevention of redevelopment) and, therefore, in situations where planning intervention is less problematical and less costly.

The historic causes of land contamination create a range of other policy complexities. These include the problem of identifying land contamination where the sources of pollution have disappeared; the related issue of deciding who should pay clean up costs where the original polluters have long gone; and the potential disruption to land values (and the economic interests connected with these) where land valuation has historically taken place with the problem of contamination unrecognized or ignored.

The legacy of past land uses is a constraint and complexity relevant to each of the forms of risk identified in Table 8.1, but to varying degrees. For nuclear facilities the problem is less central as the fact that there are risks involved has been recognized from the earliest stages of technological development. In contrast, for policies focused on reducing carbon dioxide emissions through changing patterns of transport use, the inertia of existing urban form and density is a fundamental and severe constraint (Rydin, 1995).

Identification of sources of risk

Risk identification is a key initial stage in the process of risk management. Identifying, or failing to identify, something as a potential source of risk determines whether or not attention should be given to it. It follows that for land use planning the definition of what constitutes a source of risk is crucially important in delimiting the scope of any planning action.

For major accident hazards, complexities in risk identification arise both because of the major uncertainties involved in scientific assessments of the scale of possible accidents and the fact that the storage or use of hazardous materials can be 'hidden' within outwardly innocuous activities (e.g. a water treatment works holding chlorine for water purification). Major hazards are defined in UK legislation through inventory lists of chemicals and tonnages. If, for example, an activity involved the storage of more than ten tonnes of chlorine, it would then become a notified installation and planning controls would be applied around it. This provides a clear and reasonably simple approach to identifying hazardous sites, but it is an approach that is flawed and problematical for land use planning purposes.

The uncertainty over the potential hazards of different chemicals has led to changes taking place in inventory lists and discrepancies between lists in different pieces of legislation. Changes in lists have at times been significant. For example, the original 'Seveso' Directive list has been amended three times, responding to improvements in accident modelling and the experience of actual accident events (for example, after Bhopal the threshold level for methyl isocyanate, the toxic gas released, was dramatically reduced) (Walker, 1991). Whilst such instability is perhaps inevitable in the face of scientific uncertainty, for land use planners it can be highly problematical. Local planning authorities can be faced with sites suddenly being defined as hazardous and in need of specific attention due to changes in an inventory list - or alternatively sites previously identified as significant hazards being deemed 'not so hazardous after all'. Changes in the definition of zones at risk around hazardous sites (provided to LPAs by the HSE) have been similarly problematical.

A more fundamental problem with the 'inventory list' approach to risk identification is its spatial insensitivity. Sites are defined as hazardous, or not hazardous, with no reference to their location or proximity to population. This approach to identification fails to recognise the role of location in determining vulnerability to hazards and can produce major anomalies (Walker and Draycott, 1996). For example, whilst planning controls would be applied to a site holding 25 tonnes of liquified petroleum gas (LPG) in an empty rural area, a site holding 24 tonnes in a densely populated area would not be officially identified as a source of risk, even though many more people could be affected by a fire or explosion event.

The difficulties in identifying contaminated land are similarly wrapped up in problems of scientific uncertainty, but have been more politically charged because of the economic and blight impacts of even potentially identifying a piece of land as contaminated. There is also, in most cases, no current purposeful activity which can be identified as causing the contamination (as there is, for example, with emissions from a chemical plant) and attempting to actually measure levels of contamination in sufficient detail right across the UK would be prohibitively expensive. The approach adopted to identification proposed in policy documents in the early 1990s, was to attempt to identify and target areas of potential contamination by obtaining information on past land uses (through searching old maps, registers and historical records). In 1991 a list of 42 different categories of potentially contaminating land uses was proposed that would be used by LPAs to compile registers open to public examination. However, after much protest about the scope of this list, particularly from the major financial and property interests, a much more restricted list was produced in 1992. This contained only eight categories of land use and removed an estimated 85 per cent of the land area which had been covered by the broader 1991 definition of potential contamination (DOE, 1992c). In researching reactions to the more restricted list through the analysis of responses to the DOE consultation document, the authors found that whilst some LPAs welcomed the reduction, others, particularly the urban development corporations, felt it was still too wide-ranging. The majority, however, were critical, arguing that land affected by contamination would not be identified and could therefore be built on with little or no attention given to contamination concerns.

These examples serve perhaps above all to raise doubts and uncertainties about whether or not significant sources of risk are being effectively identified and therefore coming within the remit of planning controls. Such concerns can equally be raised over, for example, the sources and levels of identified air pollution within current definitions of 'prescribed processes' and the limitations of local air quality monitoring. They also raise a problematical tension between the inevitability of short term changes in the scientific understanding and identification of risks, and the desire for stability and long term strategy that is inherent to land use planning.

Expertise and institutional boundaries

Given the technical complexities involved in estimating levels of risk, LPAs are typically reliant upon other bodies to provide expert guidance or advice on risk levels. In some cases they may be able to draw on local expertise within environmental health departments, whilst in cases where major new sources of risk are being proposed, they may be able to draw on environmental impact assessments and risk assessments produced by developers or consultants. In all cases though, statutory consultees operating at a regional or national level, such as the Environment Agency or HSE, have an important expert role to play. This has for some time been identified as a source of difficulty and ambiguity in planning control of pollution and risk (Wood, 1989; Department of the Environment, 1992b). At one level it raises basic questions about who has decision-making powers and to what extent expert advice should determine LPA actions. For major hazards the policy position has been that the HSE as the expert body has to be consulted on planning applications, but that they then provide advice to LPAs which, as locally accountable bodies, take final decisions. In practice, research has shown that it is very rare for expert advice to not be followed and there is considerable pressure on LPAs to take the expert line. DOE Circular 11/92 (1992b) makes it clear that the HSE has the right to demand the call-in of planning applications if they feel that their advice has not been properly taken into account. Where LPAs do go against expert advice, it is, in most cases, to apply lower standards than those recommended, typically through the trading-off of risk against other local concerns. It is very rare to find LPAs being able successfully to apply higher standards than those officially recommended, with any attempts to do this often obstructed by the DOE. In the case of contaminated land, the policy position is to guide LPAs to use contamination thresholds drawn up by the ICRCL (which fall below those applied in other countries such as the Netherlands) and to only require the clean up of contamination to a level that is 'suitable for use' (so that the clean up needed to construct a car park would be less than for a housing estate). Whilst there has been substantial criticism of both thresholds and clean up standards, LPAs can find it very difficult to go beyond the criteria advised by central government.

This 'tying of hands' is particularly evident for new forms of suspected hazard such as electromagnetic radiation for overhead power lines. A number of LPAs have wanted to adopt a 'precautionary approach' to this suspected hazard by introducing policies into local plans that would seek to maintain a physical buffer between new housing and overhead lines. They have been rebuffed by the DOE and the official scientific position is that no significant risk has been proven to exist (Planning, 1996).

Limitations on the scope of planning action are also seen in policy guidance on the use of planning conditions by LPAs. This guidance, for example in PPG23, strongly deters LPAs from imposing planning conditions on operational matters that are within the regulatory remit of PCAs such as the HSE or Environment Agency (DOE, 1994b). The result is that, even where LPAs have felt that other regulatory controls were inadequate or that the location of a source of risk warranted stronger than normal controls, they have either felt unable to apply additional planning conditions or found that any conditions have been rejected at appeal. However, the extent of this constraint does vary between different forms of risk, with PCAs in some circumstances actively encouraging the use of planning conditions because of authorization gaps they recognise in their own regimes.

For example, in the case of waste disposal by landfill, the use of planning conditions to monitor post-closure emissions has become the norm because the waste disposal regime itself has no authority to impose such conditions after operations have ceased at a site (DOE, 1992a). Similarly, in the case of protecting groundwater resources and reducing the risks from inadvertent spillages, it has been recognized that the land use planning system is the key instrument to control the risks posed by development, with the PCA advising on the use of conditions within planning permission. The development of general guidance notes for the protection of the water environment (National Rivers Authority, 1994) and Catchment Management Plans (Slater et al., 1994) characterise the development of a systematic approach to this process.

The importance of economic concerns

A recurrent conclusion of studies examining the practice of planning control of pollution and risk, has been that economic concerns typically dominate over environmental ones (Blowers, 1984; Wood, 1989; Miller, 1993; Healey and Shaw, 1994, Miller, 1993). Cost considerations clearly underlie a number of the issues already discussed in this chapter - for example, the reluctance to contemplate the relocation of sources of major hazard and the 'suitable for use' clean up standards applied to contaminated land. However, they are felt in a number of other important ways. The whole problematical saga of contaminated land policy has been dominated by concerns over the economic impacts of government action. A major theme of local authority responses to the 1991 and 1992 consultation papers was the cost implications - in terms of workload and resources - of implementing the contaminated land registers and the potential liability arising from incorrectly identifying sites as contaminated. Land and property owners and developers were similarly and primarily concerned with the impact of legislation on the value of their assets and the economic liabilities associated with past contamination. It would not be unreasonable in this light to see economic concerns as fundamentally obstructing policy progress in this area.

More generally there has been a concern that the expansion in the environmental remit of LPAs has been pursued with few extra resources made available from central government. This has been a particular problem in areas such as environmental risk where in order to understand sufficient of the technical and legislative complexities, training and expertise beyond that typically acquired by land use planners is needed.

Risk perceptions and access to information

Public risk perceptions can often be substantially different to technically assessed levels of risk, based upon a range of characteristics of risky technologies and the different rationalities and world views that different groups of people bring to risk issues (Royal Society, 1992). Planning can dramatically encounter different perceptions of risk in the process of policy and decision-making. Environmental pressure groups and media reports typically present very different assessments of the significance of risks to those which are conveyed through formal processes of consultation with regulatory bodies, generating sometimes intense conflict. Partly for this reason, combined with the UK tradition of official secrecy, there has been a reluctance to make information on risks freely available to the public. This has not sat easily with the more open traditions of land use planning processes.

For major hazards, lists of installations around the country were for many years considered to be confidential, this secrecy only being lifted in response to requirements in the European Community 'Seveso' Directive, implemented in

1986, for members of the public to be informed if they were living near to a designated site (Walker, 1989). The proposals for public registers of potentially contaminated land were dropped in 1993 partly because of concerns over public reactions and consequent land blight, and have not been picked up again in recent legislative proposals, maintaining a lack of public information in the face of general trends towards greater openness.

Alongside problems of secrecy, planning committees can find themselves embattled between competing interpretations of the significance of risks (Popper, 1983). Faced on the one hand by intense public concern and, on the other, by guidance in PPG 23 that LPAs should not substitute their own interpretations of risk assessments for those of the expert authorities and that 'the perception of risk should not be material to the consideration of the planning application unless the land-use consequences of such perceptions can be clearly demonstrated' (DOE, 1994b, p. 14), there are difficult political choices to be made.

Conclusion

Land use planning in the UK has, at face value, been adept at taking on board new policy agendas as the concerns and priorities of British society have evolved and developed. Healey and Shaw (1994) argue that the environment has been one area in which the 'interpretative flexibility' (p.425) of the UK planning system has been well suited to dealing with rapidly changing concerns and values. Indeed, in the case of environmental risk concerns the land use planning role has, in a relatively short space of time, become far more substantial, multi-dimensional and actively pursued. As we have identified, there is now a substantial and increasing number of environmental risks that are recognized as raising relevant planning concerns and a range of ways in which planning intervention can potentially be achieved.

A number of factors have fostered the development of this planning agenda. At the local level there have been increasing community expectations that the planning system should be a key player in local environmental policy as part of Local Agenda 21 programmes, and LPAs have, to an extent, been happy to respond by adopting a broader and more proactive approach. At a national level there have been some significant shifts towards acknowledging the importance of land use in the creation and management of risk concerns, whilst the EU has also been important in bringing forward Directives dealing with the planning-environment interface.

However, it is important to not overstate the capacity for the planning system to respond to the 'new environmentalism' (Marshall, 1994. p.21). As we have also shown, there are constraints which circumscribe the role of planning in environmental risk management, and complexities which can make policy and decision-making far from straightforward. These constraints and complexities stem in part from the limits of land use planning intervention, in part from the uncertain scientific understanding of environmental risks, and in part from the inadequacies of the policy, institutional and resource frameworks within which land use planning inputs to risk management take place.

Faced with this tension between policy potential and problems in practice, some planners have argued that they should not become embroiled in the scientific complexities of risk management. Expert regulatory bodies should instead be expected to ensure adequate levels of safety (Walker, 1994b). Blowers (1993) in contrast argues for a more central position for land use planning and advocates the integration of LPAs and regulatory bodies such as the Environment Agency, to create regional environmental planning bodies. Government policy at present is not moving down the route of further institutional integration, but rather, it is advocating greater co-ordination between planning and other agencies, for example in forthcoming requirements for the production of local air quality management plans (DOE, 1996). This may be sufficient to realise a more effective use of planning powers and undoubtedly there will be a range of situations in which LPAs can be key players in risk control - particularly where new sources of risk are being proposed and where industry translates the 1990's rhetoric of social and environmental responsibility into co-operative approaches to risk management. However, there will also be many situations where, for a range of reasons, LPAs are effectively neutered in their capacity to intervene, or where uncertain and potential environmental risk concerns fare badly against more concrete demands for development and economic gain.

References

Beck, U. (1992), *Risk Society: Towards A New Modernity*, Polity: London.

Beck, U. (1995), *Ecological Politics in an Age of Risk*, Polity: London.

Blowers, A. (1984), *Something in the Air: Corporate Power and the Environment*, Harper and Row: London.

Blowers, A. (ed.) (1993), *Planning for a Sustainable Environment*, Earthscan: London. Buckingham.

Collingridge, D. (1980), *The Social Control of Technology*, Open University Press: Milton Keynes.

Department of the Environment (1972), *Development Involving the Use or Storage in Bulk of Hazardous Materials*, Circular 1/72, HMSO: London.

Department of the Environment (1992a), *Planning, Pollution and Waste Management*, HMSO: London.

Department of the Environment (1992b), *Planning Control for Hazardous Substances*, Circular 11/92, HMSO: London.

Department of the Environment (1992c), *Environmental Protection Act 1990:*

section 143 registers, consultation on draft regulations, HMSO: London.

Department of the Environment (1994a), *Framework for Contaminated Land*, HMSO: London.

Department of the Environment (1994b), *Planning and Pollution Control*, Planning Policy Guidance Note 23, HMSO: London.

Department of the Environment (1994c) *Transport*, Planning Policy Guidance Note 13, HMSO: London.

Department of the Environment (1995), *A Guide to Risk Assessment and Risk Management for Environmental Protection*, HMSO: London.

Department of the Environment (1996), *The United Kingdom National Air Quality Strategy*, HMSO: London.

Fischoff, B., Hohenemeser C., Kasperson, R. and Kates, R. (1978), Handling Hazards, *Environment*, Vol.20, No. 7, pp. 161-79.

Healey, P. and Shaw, T. (1994), Changing Meanings of Environment in the British Planning System, *Transactions of the Institute of British Geographers*, Vol. 19, No. 4, pp. 425-38.

Health and Safety Executive (1983), *The Fire and Explosions at B&R Hauliers, Salford*, HMSO: London.

Health and Safety Executive (1989), *Risk Criteria for Land Use Planning in the Vicinity of Major Industrial Hazards*, HMSO: London.

Health and Safety Executive (1993), *The Fire at Allied Colloids Limited*, Health and Safety Executive: London.

Her Majestys Inspectorate of Pollution (1995), *Planning Liaison with Local Authorities*, HMIP: London.

Interdepartmental Liaison Group on Risk Assessment (1995), *The Use of Risk Assessment within Government Departments*, Health and Safety Executive: London.

Lagadec, P. (1982), *Major Technological Risk*, Pergamon: Oxford.

Marshall, T. (1994), British Planning and the New Environmentalism, *Planning Practice and Research*, Vol. 9, No. 1, pp. 21-30.

Miller, C. (1993), 'Coke, Smoke and National Sovereignty: A Case Study of the Role of Planning in Controlling Pollution', *Journal of Environmental Planning and Management*, Vol. 36, No. 2, pp. 149-66.

Miller, C. and Fricker, C. (1993), 'Planning and Hazard', *Progress in Planning*, No. 40, pp. 167-260.

National Rivers Authority (1994), *Contaminated Land and the Water Environment*, NRA: Bristol.

O'Riordan, T. (1979), The Scope of Environmental Risk Management, *Ambio*, Vol. 8, No. 6, pp 260-4.

Owens, S. (1994), 'Land, Limits and Sustainability: A Conceptual Framework and Some Dilemmas for the Planning System', *Transactions of the Institute of British Geographers*, Vol. 19, No. 4, pp. 439-56.

Petts, J. (1988), 'Planning and Hazardous Installation Control', *Progress in Planning*, 29, pp. 1-75.

Petts, J. (1995), 'Contaminated Sites: Blight, Public Concerns and Communication', *Land Contamination and Remediation*, Vol.2, No.4, 171-81.

Planning (1996), 'Bristol Research Suggests Link From Power Cables to Cancers', *Planning*, No. 1156, pp. 7.

Popper, F.J. (1983), 'LP/HC and Lulus: the Political Uses of Risk Analysis in Land Use Planning', *Risk Analysis*, Vol. 3, No. 4, pp. 255-63.

Royal Society (1992), *Risk Analysis, Perception, Management*, Royal Society: London.

Rydin, Y. (1995), 'Sustainable Development and Land Use Planning', *Area*, Vol. 27, No. 4, pp. 369-77.

Slater, S. Marvin, S. and Newson, M. (1994), 'Land Use Planning and the Water Sector'. *Town Planning Review*, Vol. 65, No. 4, pp. 375-97.

Talbot, J. (1996), 'Defining Contaminated Land - A Crisis in the Making', *Town and Country Planning*, Vol. 65, No. 9, pp.238-40.

Walker, G. P. and Macgill, S. M. (1985), 'Planning Control of Industrial Hazard in a Major Metropolitan Planning Authority', *Environment and Planning B*, Vol. 12, pp. 463-78.

Walker, G. P. (1984), 'Case Study: Planning Applications and Major Hazards', in Petts, J. (ed.), *Major Hazard Installations: Planning Implications and Problems*, Department of Chemical Engineering, Loughborough University of Technology: Loughborough.

Walker, G. P. (1989), 'Risks, Rights and Secrets: Public Access to Information on Industrial Major Hazards', *Policy and Politics*, Vol. 17, No.3, pp. 255-71.

Walker, G. P. (1991), 'Land Use Planning and Industrial Hazards: A Role for the European Community', *Land Use Policy*, Vol. 8, No. 3, pp. 227-40.

Walker, G. P. (1994a), 'Industrial Hazards, Vulnerability and Planning in Third World Cities', in Main, H. and Williams, S.W. (eds), *Environment and Housing in Third World Cities*, Belhaven: London.

Walker, G. P. (1994b), *Hazardous Substances Consents: A Review of the Operation of Statutory Planning Controls Over Hazardous Substances*, Department of Environment: London.

Walker, G. P. (1995), 'Land Use Planning, Industrial Hazards and the 'COMAH' Directive', *Land Use Policy*, Vol. 12, No. 3, pp. 187-91.

Walker, G.P. and Bayliss, D. (1997), 'Development Plans and Hazardous Installations', *Planning Practice and Research*, Vol.12, No.4.

Walker, G. P. and Draycott, P. (1995), 'Mapping Technological Hazards: Spatial Patterns of Major Accident Hazards in England', *Occasional Paper New Series A*, No. 7, Division of Geography, Staffordshire University: Stoke on Trent.

Wood, C. (1989), *Planning Pollution Prevention*, Heinemann: London.

9 Derelict land - some positive perspectives

Philip Kivell and Sarah Hatfield

Introduction

In many European countries the economic restructuring of the past three decades has produced a legacy of extensive industrial dereliction. This is especially widespread in the cities that prospered during the nineteenth century manufacturing boom. The result is that many of these cities are characterized by expanses of derelict land, together with smaller unused and vacant sites, that occur in a physical state and in locations that are unattractive to developers today. In England alone, the most recent survey (DOE, 1995) estimated that a total of 39,600 hectares were derelict in 1993, and that, despite major reclamation programmes, the total had decreased by less than 10 per cent in the previous twenty years. In fact the scale of the problem is undoubtedly larger than these figures imply. Because of the way in which the figures are collected, and the way in which the definitions are drawn, there is a widely held view that the real total is much higher. Some estimates have put the total of derelict and vacant land in England as high as 200,000 hectares (Chisholm and Kivell, 1987).

The conventional perception of derelict land is almost wholly negative, and there are very good and powerful reasons for this. Essentially this view can be justified on three counts. First, derelict land represents a waste of a valuable resource. Britain is a small and congested country so there are both economic and moral arguments for not allowing land to be wasted or to stand idle. During the 1980s and 1990s planners and politicians have faced the dilemma of having to find development land, often on the edges of cities, whilst large numbers of derelict sites remained unwanted within inner city localities. During the development-led planning era of the 1980s, any unused land that was not coming forward for development, or was not producing revenue, was seen to be an economic and planning liability.

Second, derelict land can be ugly. By reason of the industrial activities that it originally supported, and the nature of its abandonment, much derelict land is scarred and damaged. Although natural regeneration can eventually cover many of the scars, land that is damaged and unused often attracts litter, tipping, and vermin and can become associated with anti social or illegal activities, all of which make it even more unattractive. Many of these problems spread beyond the immediate boundaries of the site and serve to blight the local environment, damage the confidence of the local community, break up the harmony of the townscape and detract potential investment in the wider area. Third, much derelict land is dangerous. Previous mining activities can create unstable tips, shafts, pits (often flooded) and voids, and industrial activities frequently leave behind dangerous buildings and equipment or toxic contamination.

The negative aspects of derelict land are emphasized in the definition used by the Department of the Environment and local planners, i.e. 'land so damaged by industrial, or other development that it is incapable of beneficial use without further treatment' (DOE, 1995, p.3). The public view of dereliction has been largely shaped by the negative images portrayed in the media (see for example Atkinson, 1993; Miller and Warren, 1993; Walsh, 1991). In recent years both television and the newspapers have used pictures of derelict land to illustrate issues as diverse as the loss of industrial jobs in the North East, the rising drug and crime problem in parts of Manchester and the political competition for votes in the inner city. Against this background it is not surprising that the overwhelming perceptions of derelict land, amongst planners, politicians and the general public, are negative ones.

It is not the purpose of this chapter to deny the validity of these negative images, nor to argue for the wholesale retention of derelict sites. However, by recognising that derelict land is, and will continue for many years to be, a fact of life for thousands of city dwellers, the authors attempt to present a complementary view by suggesting that it can also have identifiable benefits and positive attributes. The argument is located largely within a study of North Staffordshire, based upon research carried out in the period between 1993-1996.

The detailed survey was conducted upon five relatively large derelict sites (see Figure 9.1), which were chosen in order to provide a range of different kinds and stages of dereliction. After a series of pilot questionnaires and initial observations of all sites, more detailed questionnaire surveys were undertaken in a sample of residential areas and schools adjacent to the sites.

Before examining the positive facets of derelict land, it is necessary to consider the way in which local people conceived and defined the nature of dereliction. All of the sites surveyed fell within the planner's definitions of derelict, but the views of local people varied greatly. Whereas the formal definition relies largely upon economic criteria and relatively narrow considerations of 'beneficial' use, the views of the local communities were more subjective. When asked whether they

Figure 9.1 Derelict sites surveyed in North Staffordshire

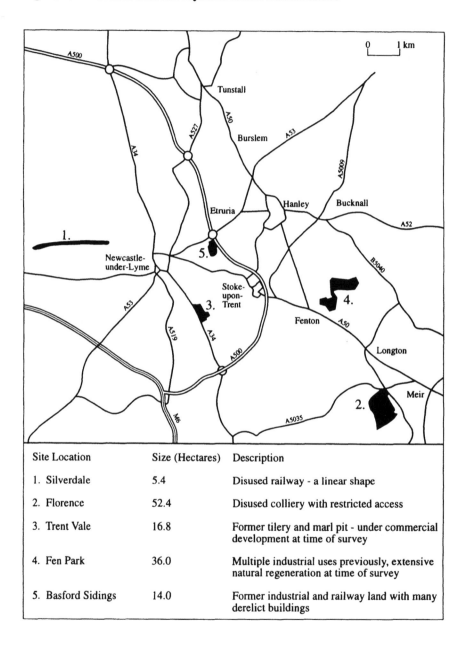

Site Location	Size (Hectares)	Description
1. Silverdale	5.4	Disused railway - a linear shape
2. Florence	52.4	Disused colliery with restricted access
3. Trent Vale	16.8	Former tilery and marl pit - under commercial development at time of survey
4. Fen Park	36.0	Multiple industrial uses previously, extensive natural regeneration at time of survey
5. Basford Sidings	14.0	Former industrial and railway land with many derelict buildings

considered the five sites in question to be derelict, local responses ranged from 76 per cent in agreement at the Silverdale site to only 20 per cent in the case of Fenpark. Many respondents agreed that the respective sites were derelict (a term which has widespread if imprecise, local currency), but not useless. It is clear that in the minds of many people there was an unresolved dichotomy. On the one hand a site was derelict because former economic activity had ceased and/or it was untidy, but on the other hand it was not useless because some natural regeneration had taken place and people used it unofficially for a variety of informal activities. A number of studies have shown that derelict sites have positive value. For example in a survey conducted among amenity societies and community groups by the Civic Trust (1988), 59 per cent of respondents said that there were areas of wasteland that were assets and the same study referred to 40 per cent of all derelict sites in Stoke on Trent as having some informal use. In the present study, 63 per cent of respondents, when first asked, suggested that there were some advantages in living adjacent to the derelict sites with various forms of recreation and wildlife interest accounting for almost two-thirds of the specific factors named.

Most of the rest of this chapter will be concerned with the positive values and functions that local residents attributed to these officially derelict sites, and with an attempt to outline the main factors that influence those views. In some cases a positive reaction to derelict space can be seen to stem from relatively intangible attributes, such as its aesthetic quality, or the historical and cultural heritage it symbolises, in other cases the value derives from a more direct set of uses and activities especially in satisfying recreational needs.

The use of derelict sites

Aesthetics and landscape

Geographers have paid considerable attention to landscape studies in recent years (see for example Appleton, 1996; Cosgrove and Daniels, 1988; Crouch, 1993; Meinig, 1979). Even though many of these have examined 'ordinary' landscapes, there has been practically no mention of derelict areas. In his seminal work *Topophilia*, Tuan (1974) largely ignored such blighted spaces, perhaps implying that nobody loves them.

Despite the overwhelmingly negative view of derelict land portrayed by the media and official planning documentation, there is evidence from the present study that many people perceive it to have some positive aesthetic attributes. In particular, derelict land that had a relatively open aspect was seen by many as preferable to continuous buildings and it was valued for it role in breaking up a high density urban landscape. Even where people agreed that it was not aesthetically attractive per se, they often referred to some redeeming features such as the presence of greenery or wildlife. Although there was much duality of response, reflecting ambivalent attitudes, the community views on the whole were more favourable than unfavourable. The length of time that an individual site had been derelict also had some bearing upon people's views, especially with respect to the amount of natural regeneration that had taken place through the growth of grass and shrubs. In a city such as Stoke on Trent, where the prevailing type of urban development is undistinguished, greenery, even if it is on 'derelict' sites, is clearly valued. Respondents commented that even when sites only appeared green from a distance, this enhanced the appearance of the environment. It was not only the natural greening of sites that attracted positive perceptions; some respondents found that decaying Victorian industrial buildings provided a stark beauty of their own.

At the one site, in Trent Vale, where commercial development was planned at the time of the survey, the chief objection to the new use as an out of town retail park was based upon aesthetic grounds and the destruction of a semi- natural green landscape.

Here it is worth noting the difference outlined by Nohl (1985) between perceptive cognition and symptomatic cognition, because respondents in the present survey, probably unknowingly, distinguished between them. At the level of perceptive cognition people commented upon, for example, the green-ness of grass and shrubs and the informality of the landforms. Beyond this however, at the level of Nohl's symptomatic cognition, or what others have referred to as associational meanings of environment, respondents recognized that elements in the landscape also pointed to uses for the open space. Thus the green open spaces afforded an opportunity to escape the congested built-up city and the nature of the land invited activities, such as riding motor cycles or walking dogs, which might be discouraged on more formally managed open space. There is also evidence from the survey that age has an important bearing upon aesthetic judgements. Older age groups were more concerned with the appearance of the sites, but in the sample of school children aesthetic issues were relatively unimportant. What mattered to them was the function of the site and the possibilities it afforded for active use. This point will be returned to later.

Community space

In the vacuum created by the decline of traditional industrial land uses, and the inability of planners and politicians fully to cope with these changes, there has been a tendency for some communities to take matters into their own hands. Many years ago the National Council for Voluntary Organisations published a guide for community groups to help them use urban wasteland (Lobbenberg, 1981) and later, Shirley-Smith (1989) showed that in the European Year of the Environment there was widespread popular action to re-occupy derelict land in many of Europe's inner city areas. People from all levels of society including children, families, retired people and recent immigrants began to occupy or appropriate derelict and vacant sites for a range of purposes including recreation, gardening and wildlife preservation. Sometimes this was done with the active or tacit agreement of the planners, but more often it was informal, spontaneous community action in the face of apathy or inactivity on the part of landowners and official bodies.

In the present study there was a very real and widespread sense in which local residents considered the derelict sites to be an integral part of their community. This was revealed both by the extent and variety of local uses and by the community opposition to certain redevelopment proposals. A sample of residents living within an arbitrarily defined zone within 750 metres of each of the five sites surveyed revealed that 75 per cent of respondents made some use of them. Over half of the respondents (53 per cent) reported regular use on a least a weekly basis. Systematic observation revealed a wide variety of informal recreational activities and the fact that the sites are considered real community assets was reinforced by additional comments from many respondents. Much of the activity was individual, or was carried out in small groups, but some of it provided a genuine community focus, as with Bonfire Parties on November 5th and community organized summer activities for children at Fen Park and Basford, for example. As other community spaces decline, or become increasingly regulated, or become subject to increasing dangers from motor vehicles, the relatively uncongested and uncontrolled nature of derelict sites gives them a particular value which can be utilized for the needs of the local community, including those activities which do not conform to more formal codes of practise.

Such sites also offer particular advantages to local communities in terms of access, proximity and cost. The majority of the uses recorded were extremely localized, thus accentuating the community or neighbourhood value. Proximity to home was seen to be important, but proximity to place of work, or to school was significant, especially given the high level of use recorded by children and teenagers. For these groups in particular, the absence of any direct financial cost was important, especially since many of the more formal leisure sites have recently become more commercialized, and more costly to use. Conditions of access affect the potential for use and it was noticeable that at the Florence site, where mining use had only recently ceased and there was thought to be a strong likelihood of commercial redevelopment, the site was largely inaccessible for community use because of elaborate fencing and security arrangements. Conditions of access also helped to determine who might use the site and the times of the day or week when use was possible. In this respect teenagers proved notably more adventurous than other groups of the local population. Teenagers, more than any other groups, made use of the sites after dark thus, in effect, reclaiming time as well as space for their own use.

The value that local communities placed upon the derelict sites was also revealed by their responses to a question about their preferences regarding future development. Communities appeared to become even more keenly aware of the value of derelict sites when there was a proposal to develop them, a finding also noted by Freeman (1992) working in Leeds. In summary, the preferences stated by respondents in the present survey were:- 1. To leave the site as it is, 2. Tidy it up, 3. Make a recreational area/children's play area, 4. Make a wildlife park, 5. Not to build upon it. Opposition to formal development was based upon loss of recreational area and wildlife, loss of green space or outlook and loss of history/ heritage, but there was also some reference to the possibility of deviant behaviour relocating away from the derelict sites and back into residential neighbourhoods. In the case of any proposed development, the importance of consultation with the local community over the future of the areas was stressed by many respondents, and there was a clear feeling that they currently had insufficient power to influence planning decisions.

Heritage

Closely connected with community aspects of local land use are issues relating to heritage. In the case of derelict land there is a very direct link with the past in that the dereliction has often been created by the decline of traditional heavy industries upon which the economic and social life of the community had depended. In this area there has long been a dependence upon a few industries, notably pottery manufacturing and coal mining, and a history of family employment in the same factory or colliery over several generations. Lewis (1979) suggested that the human landscape forms a part of our unwitting biography, reflecting our tastes, values, aspirations, and even our fears, in a tangible form. Even derelict landscapes play a role in this process, representing a clear part of the identity of both individual's and communities' lives.

The notion of heritage is often a very personal one in which people grow accustomed to their familiar environments and the symbolic values that they possess. Industrial landscapes can hold a fascination, and even a beauty of their own, especially to people who have a long history of dependence upon those industries. This became clear from the comments made by many of the respondents. On most of the sites in question, the more obvious industrial relics have already disappeared; railway lines have been removed, pit head gear has been demolished and buildings have been razed, but the sites retain a strong heritage value and have important meaning for local people. In recognition of this, some respondents made reference to the possibility of heritage museums or local history trails as alternative uses for derelict land. Obviously there is a limit to the number of such projects that can be sustained and it brings into question the whole, delicate, concept of heritage. We are certainly aware, as were many of the local residents, that there are at least two inherent problems. Firstly the introduction of elaborate preservation measures can have the effect of formalising an informal space, thus altering its nature, use and accessibility. Secondly there is the possibility that Heritage Projects can result in stage-managed, even phoney, developments that compromise the very nature of heritage itself. At the least, what we are suggesting from the evidence of this survey is that much derelict land contains strong, albeit subtle, heritage value in a relatively unmanipulated form, and that for this reason the conventional image of dereliction, as always undesirable and needing eradication, must be approached with caution.

Recreation

The foregoing has identified a number of positive values and uses that derelict land possesses, but it is clear that these, although important, are also quite broad, and sometimes hard to pin down. If we focus upon observed and reported activities in a more tangible sense, it is recreation, broadly defined, which is most frequently served by derelict sites. In the community questionnaire, 59 per cent of respondents reported that active recreation was their main use for the derelict sites (this excludes watching wildlife).

Recreation covers a very wide range of activities; amongst those observed in this survey were, informal play by children, hide and seek, spontaneous team games (including football and cricket), golf, riding bicycles and motor bicycles, walking (alone or with a dog), jogging, horse riding and meeting friends. Local residents had even stocked two flooded marl pits with fish to enable informal coarse fishing to take place. The very informality of these activities on derelict sites invariable means that they defy easy categorization. The reasons why derelict land is so used are fairly straightforward and many of them, including proximity, access and cost have been alluded to above. There are, however, other important reasons. Among these is the lack of alternative space. The original high density form of development, which still characterises much of the local urban area, made little provision for recreational space, either within the individual home and garden or within the neighbourhood. Much of the land which has subsequently become available has been used for roads, car parks or institutional use, so people, especially young people, resort to derelict and vacant land. Other attractions include unusual topographies or installations on the sites which may offer challenges to more adventurous play or motor cycling.

Finally, and most significantly for some, derelict sites often provide a freedom of use which does not apply elsewhere. Such sites are not directly controlled or supervised, they are not subject to rules of use or opening hours and there is less likelihood of complaint or conflict with other, more formal, activities. Obviously there is a danger here that such freedoms may also permit undesirable, anti-social or deviant behaviour. There was some evidence of this in the form of under-age drinking, smoking, and sex, substance abuse and the use of unlicensed motor bikes. Some of the respondents were well aware of this and even made the point that such activities were inevitable, but at least derelict land provided an arena for them to take place away from localities where they would cause greater offence.

People of all ages reported using derelict land for recreation, but their motivations, activities and even their times of use varied greatly. For elderly people the convenience, or proximity of some derelict sites to their homes was a clear attraction, especially where they were unable through infirmity or lack of transport to travel further. Above all, however, it was clear that children and teenagers were the most extensive users of the sites in the survey. A number of previous studies (Civic Trust, 1988; CoEnCo,1981) have shown that derelict sites often have interestingly varied appearances and topographies which, together with the 'forbidden' aspect, can make them compelling venues for more adventurous play. This was certainly borne out by the present survey and some respondents alluded to the satisfaction to be gained from invading the derelict space and claiming it for their own. We would also agree with Woodward (1990) and Freeman (1992) that innovative, unregulated play on what is effectively 'unprogrammed space' can be an important part of a child's learning process, helping him/her to develop necessary life skills.

It was found that teenagers used derelict land even more than children; indeed teenagers used derelict sites more frequently, for longer periods of time and for more different activities than any other group. Teenagers who responded to the survey conducted in local schools showed that they were less concerned than other people with the appearance of derelict sites and more concerned with its immediate potential for use. For teenagers, the main attraction of such sites was the freedom from adult supervision and control that they afforded. This was exploited to the full with teenagers using the more hidden and remote parts of the sites and also using them at times of the day (especially evenings) when conflict with other users and the constraints of supervision could best be avoided. At the extreme this permitted anti-social, even illegal activities to take place. The other factor helping to explain the attraction for teenagers is the lack of alternative recreational spaces for them. They are (officially) too young for pubs, too old for what they see as the tame activities and spaces provided for younger children, and have insufficient money to visit commercial leisure activities more than occasionally.

Derelict land and nature

Because of the intensive, and orderly way, in which much urban open space is managed, flora and fauna often have relatively little chance to develop naturally. However, upon derelict and other vacant sites nature often has more of a free rein and some quite rare species can thrive. Observing flora and fauna can, in one sense, be regarded as a recreational activity, but the significance of wildlife in the city also goes far beyond this, as has been confirmed by Harrison, Limb and Burgess (1987) inter alia. Essentially they argued that encounters with nature and wildlife in the city confer personal and social benefits for the individual by creating a sense of fun, marvel and delight, by satisfying a desire for adventure and by providing variety from over-institutionalized landscapes. Nature, wildlife and open spaces help to provide 'gateways to a better world'. In recent years the tentative greening of popular culture and a growing environmental awareness have enhanced this aspect of urban living and given people a greater appreciation of nature.

Almost two-thirds of adults in the present survey reported that they used derelict land as an area for observing wildlife and it was listed as one of the most commonly perceived advantages of living close to derelict land. Although the majority of species present are not particularly rare, people reported gaining pleasure from them and were ready to list the animals, birds, insects and plants that they had seen. Even where individual aspects were not singled out for praise respondents were keen to point out the links between nature, a green appearance and the semblance of countryside in the city. Respondents in all age groups made some reference to the wildlife and natural aspects of derelict land, but it was particularly important for retired people, many of whom were not physically able to involve themselves in active recreational pursuits. Older members of the community referred to watching wildlife from their homes, or from the edges of the sites, and many of them expressed pleasure at the way in which wildlife actually spread outside of the boundaries of the derelict sites into the adjacent neighbourhood. The lack of active land-use management, and the absence of pesticides and other controls also meant that derelict sites provided valuable ecological niches unavailable elsewhere in the city.

Planning

There is no doubt that the over riding ethos within planning circles sees derelict land as being a problem which needs solution and the development pressure upon planning in the 1980s and 1990s has emphasized the need to redevelop such sites. Partly this has been driven by the need to provide economically beneficial uses such as housing or commercial activities, but equally important has been the intention to improve the quality of the urban environment by removing eyesores. This in turn has had the dual aims of improving the appearance of areas for their local communities, and making them more attractive to inward investment. Whilst these aims are entirely understandable, it is worth recognising that economic and social needs do not always point to the same solution and that, derelict sites may well offer community benefits without formal development. In particular respondents to the present survey felt that they often had little say in how local planning decision were taken, and that planning solutions were imposed from outside of the community. This prompts questions about community identity, the ways in which places are formed from space, how that space is contested and who has power to influence events.

Perhaps there is a case for suggesting that a highly structured, modernist approach to land use planning has produced over planned, institutionalized spaces. This is particularly true of open spaces such as playing fields, public parks and the open space around housing. Nature, and even some of the human needs for freedom and spontaneity are being denied by design and planners are insufficiently aware of the deeper meanings that people attach to space in their neighbourhood. As Nohl (1985) argued, existing, planned open spaces often lack variety, local individuality and historical context. Formalized open spaces are presented as valuable and unalterable works of art, but they remain alien to many people. Only when we can experience landscape through our own actions can they become meaningful living spaces with personal associations. Perhaps in the emerging paradigm of the post-industrial city, with less emphasis upon formal, tidy and highly ordered elements, and more emphasis upon flexibility and variety, there will be a greater appreciation of the value of derelict spaces in their undeveloped form.

Conclusion

Whilst dereliction can undoubtedly be the source of many problems, the argument presented here is that it also possesses some positive aspects which are valued by local communities. There is evidence to suggest that local communities, or at least some elements within them, notably young people, interpret and evaluate the characteristics of derelict land differently from planners and other official agencies. The idea that derelict land is economically unproductive, and therefore bad is an oversimplification. Harrison et al. (1987, p.360) found that '...the left-over pockets of waste ground and derelict areas, with their happy mix of wildness and remains of buildings, hold an important place in people's lives', and we concur with this view. In some cases there was a clear sense in which people had effectively 'colonized' derelict sites within their neighbourhoods and were deriving a range of benefits from them. It could be argued that an appreciation of derelict land is simply making the best of a bad job, but the evidence from this survey suggests that some derelict sites offer very real and important benefits, most usually in communities where other opportunities are rather poor. Some of these benefits are difficult to quantify in an objective manner, being concerned with aesthetic qualities, symbolic values and heritage, but others, such as recreational functions and wildlife habitats, can be identified in much more tangible forms. With the renewed emphasis now being placed upon the contribution that derelict land can make towards the provision of sites for new housing over the next twenty years, local communities may well find that their derelict sites are attracting new interest from developers.

References

Appleton, J. (1996), *The Experience of Landscape*, Wiley: Chichester.

Atkinson, D. (1993), 'Paying the Price', *Impact*, (Quarterly magazine of the Guardian), Autumn, London.

Chisholm, M. and Kivell, P.T. (1987), *Inner City Wasteland*, Institute of Economic Affairs: London.

Civic Trust. (1988), *Urban Wasteland Now*, Civic Trust: London.

CoEnCo. (1981), *Waking up Dormant Land*, Council for Environmental Conservation: London.

Cosgrove, D. and Daniels, S. (eds) (1988), *The Iconography of Landscape*, Cambridge University Press: Cambridge.

Crouch, D. (1993), 'Representing Ourselves in the Landscape', in Brown, R. (ed.) *Continuities in Popular Culture*, Popular Press: Ohio.

DOE. (1995), *Survey of Derelict Land in England, 1993*, Department of the Environment, HMSO: London.

Freeman, C. (1992), *Under Threat? Naturally Regenerating Wasteland in Leeds*, CUDEM Working Paper No. 15, University of Leeds: Leeds.

Harrison, C., Limb, M. and Burgess, J. (1987), 'Nature in the City - Popular Values for a Living World', *Journal of Environmental Management*, Vol. 25, pp. 347-62.

Lewis, P. F. (1979), 'Axioms for Reading the Landscape', in Meinig, D.W. (ed.) *The Interpretation of Ordinary Landscapes*, pp. 11-33, Oxford University Press: New York.

Lobbenberg, S. (1981), *Using Urban Wasteland*, National Council for Voluntary Organisations: London.

Meinig, D.W. (ed.) (1979), *The Interpretation of Ordinary Landscapes*, Oxford University Press: New York.

Miller, R. and Warren, P. (1993), 'A Cry from the Streets', *Sunday Times Magazine*, March 21st. London.

Nohl, W. (1985), 'Open Space in Cities: Inventing a New Esthetic', *Landscape*, Vol. 28, Part 2, pp. 35-40.

Shirley-Smith, C. (1989), 'A Patchwork View of Action Involving Community Groups on Ten Derelict Land Sites in Europe', *Landscape Research*, Vol. 14, Part 2, pp. 11-17.

Tuan, Y-F. (1974), *Topophilia*, Prentice-Hall: Englewood Cliffs.

Walsh, J. (1991), 'Isle of Despair', *Time Magazine*, March 15th.

Woodward, S.C. (1990), *The Phenomenon of Vacant Land in Stoke-on-Trent*, Unpublished Ph.D., Staffordshire Polytechnic.

10 Sustainable development and spatial reorganization of a Greek border village with the help of private capital

Evangelos Dimitriadis, Alexandros Ph. Lagopoulos and Christos Th. Kousidonis

Introduction

This chapter consists of two parts: the first part traces briefly the history of planning in modern Greece, and the second investigates the case of a mountainous border settlement in Northern Greece called Archangelos.

Greece: historical retrospective on planning, land use and the environment

The history of modern Greece starts with the liberation from the Turks in 1827 and the establishment of the modern Greek state. This was a gradual process, starting from the geographical areas of the Peloponnese and Sterea Hellas (1832), followed by the liberation of Thessaly in 1881, and finally of Macedonia in 1912-13 and Thrace in 1923.

The modern history of the country can be divided into three periods (A, B, C), which are characterized by the gradually more intense presence of capitalism, which spread from south to north following the parallel extension of the national boundaries (Lagopoulos, 1992). It is important to analyse these three periods in order to understand both the socioeconomic mechanisms of the development of the country, and the applied design practice which was inherited by the contemporary Greek state (Figure 10.1).

First period

The first period A (1827-1907/13) witnessed the beginning of the transformation of Greece from a dependent feudal region of the collapsed Ottoman Empire to a modern urbanized state, with institutions of a western European type and an economy increasingly controlled by capital. The same period is characterized by urbanization and the development of new social forces. This first period was dominated by merchantile capitalism. It can be divided into three subperiods: A1, from 1827 to 1844; A2, from 1845 to 1859; and A3, from 1860 to 1913.

Figure 10.1   Historical development of planning in Greece

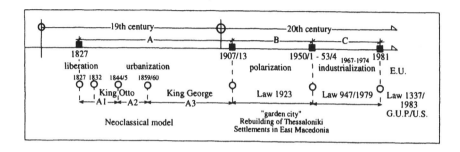

In the first decades of the new state (A1), the neoclassical model of a city was introduced as an innovation by the foreign technicians and architects who worked in Greece until 1833. This urban model served the bourgeoisie and the royal court (with its largely foreign culture), because it secured the connection with the West; it symbolized the rejection of the immediate Ottoman past and attempted to connect the modern kingdom of the Bavarian King Otto (1833-43/62) with the ancient Greek world (see Kafkoula et al., 1990). During this first subperiod plans were drawn up for 25 cities.

In the second subperiod (A2), plans of nine towns were produced during an era of wide political and economic instability, when the foreign technicians withdrew and were replaced by Greek ones with less technical training. During this period planning models were redirected towards realistic applications adjusted to Greek conditions.

Finally, the third subperiod (A3) introduced a state mechanism for planning, with an integrated framework of institutions and legislation administered by special agencies. The original neoclassical urban model was applied, though in impoverished form, even to small settlements. The rate of production of new city plans stabilized at about five new plans per year.

The importance of urban planning in the first period, especially of the central settlements of the new nation, was twofold. On the one hand, it acted as a means to establish the sovereignty of the modern state in the context of a renovated network of settlements. On the other, it responded to the real need for housing of populations living in settlements destroyed by war.

In conclusion, this first period, which covered almost the whole of the 19th century, witnessed an attempted innovative reorganization of Greek national space, emphasizing urbanization in a country which was clearly predominantly rural. By 1845 plans had been drawn up for almost all the important centres of the country (what is today southern Greece), with priority given to the regional harbours and the settlements important in the administrative hierarchy. By 1879, plans had been produced for 93 settlements (with between 5,000 and 25,000 inhabitants). Finally, by 1912, 174 settlements (from 500 to 20,000 inhabitants) had been given a new plan. The modern Greek city, regardless of size, had become part of an integrated system of public administration which was aimed at creating a homogenous urban space according to the neoclassical principles already mentioned.

Second period

The second period (B) represents the first stage of capitalism: liberal capitalism. This period was characterized by political instability (two world wars and one civil war) and the depopulation of the country. Economic growth was accompanied by an increase in the degree of hierarchy and complexity of the settlement network, especially after 1920. We can observe a continuously intensified socioeconomic and demographic polarization among the settlements, accompanied by intensified urbanization. The main characteristics of the settlement network were the polarization between Athens/Piraeus and Thessaloniki, and the creation of an axis of development along the road link between Athens and Thessaloniki.

In 1923 and the immediately following years, this gradually emerging national settlement network received a massive influx of more than one million refugees from Asia Minor, who settled with no preparation mainly in the two basic poles of the system. This, in conjunction with other important political events, led to a serious lack of organization of land-use as well as to the downgrading of the physical environment.

However, in the first quarter of the 20th century, there were also efforts (by the sociologist Minister of Transport and Communications A. Papanastasiou) to introduce Ebenezer Howard's model of the garden-city into Greek design practice, as part of the modernizing policy of the liberal government of Eleftherios Venizelos.

This model emerges in two cases of planning in Northern Greece, firstly, with the rebuilding of Thessaloniki after the fire of 1917 (Kalogirou, 1983), and secondly, with the planning of a group of settlements in East Macedonia destroyed by the war (Kafkoula, 1990). For these two parallel cases two international groups of planners and architects were assembled, the first one (1918) headed by Ernest Hebrard (Lavedan, 1921) and the second (1919-20) by the English architect John Mawson, son of the city planner Thomas Mawson who had worked with Hebrard's group on Thessaloniki.

It seems, however, that the model of the garden-city was not compatible with the existing problems of Greek society in the beginning of the 20th century. Thus, with few exceptions - for example, Pl. Dracoulis, who in 1924 founded the Greek Society of Garden Cities, and the plans of the neighborhoods of Psychiko (1923) and Ekali (1924) in Athens - the garden-city model was not incorporated into the planning system.

During the 20th century the housing problem began to dominate planning, especially in the urban areas. Gradually, the settlements of the rural regions were depopulated, especially in the border areas both insular and inland; by the 1970s there were approximately 300,000 uninhabited houses in these areas. In the urban centres the exploitation of land became intense and large areas of illegally constructed housing appeared, which were shamelessly exploited by the building contractors. Similarly, in the city centres (C.B.D.) development was left to private initiative without any general planning. This resulted in a high concentration of economic and administrative services in the central areas of the large urban centres, accompanied by the destruction of the historic urban environment, inefficient functional and transport management and a lack of infrastructure (Lagopoulos, 1992).

Industry was located with no planning in and around the cities, where cheap labour was available, without suitable infrastructure; this caused a serious pollution of the physical environment (KEPE, 1976; TEE, 1979).

Attempts by the government to respond to planning problems became more substantial toward the middle of the first period (A). These took the form of investment in transportation infrastructure at the regional level. In practice, planning still mainly involved city plans, in the neoclassical style which also had an ideological role to play. The second period (B) saw a rationalization of state spatial planning, at least on a theoretical level. A policy which supported the formulation of fragmentary plans and small public works, supported by a basic law of 1923, can also be dated to this period. The main problem was the settlement of refugees, in which small private capital played the most important role (Lagopoulos, 1992).

Third period

The third period (C: 1951/4-1981) corresponds to the development of monopolistic capitalism. A very small native bourgeoisie controlled local capital, which at the same time was itself dependent on international financial and political centres. These centres supported the military dictatorship of 1967-74. Demographically the lower-middle class was predominant, while there was also a significant working-class: in 1971, manual labourers constituted 30 per cent of the working population.

After 1960 Greece modernized according to the capitalist model of development, based primarily on a rapid rate of industrialization, while the economy was controlled by European and American capital (Benas, 1978; Svoronos, 1964). Greece's accession to the European Community in 1981 expanded the southern part of the Community and set the conditions for closer relations between the Community and the countries of the Eastern Mediterranean and North Africa. Greece reoriented itself to act in concert with the other countries of the European Union on issues such as the internal market, currency, economic and social cohesion, as well as foreign policy and security (Fontaine, 1995).

The manner of development of the country in these decades intensified the phenomena already noted above. Housing, with the exception of a very few public projects, was left in the hands of private capital and was carried out on the basis of the construction of single buildings and not housing areas. Physical planning was rationalized and specialized, but it was still not generally applied. At an administrative level, state intervention increased, manifested through a bureaucratic administrative machinery and through legislation. Starting in 1960, regional and master plans were drawn up but not implemented, while state intervention was usually limited, fragmentary and characterized by wishful thinking (TEE, 1978).

Urban legislation developed mainly during the 1970s. The new constitution of 1975 provided a general framework for urban, regional and environmental policy (Lagopoulos, 1992). The state was responsible for the planning of the economy and regional development, as well as for the development and protection of the physical and cultural environment. For the first time there was constitutional recognition of the social responsibility of private property, the right to which cannot take priority over the public interest. Legislation introduced the concept of the development area, in which urban development can be financed through the participation of local property owners contributing in money or land.

Further legislation defined the concepts of the regional, master and urban plan, the control area around the master plan, and the concept of land use (Law 1262/1972). In other legislation, the concept of the development area was specified and the concepts of general and specific land uses introduced, with the city block being used as the smallest geographical unit (Law 947/1979; Christophilopoulos, 1979). This legislation was an attempt at modernization in view of the country's accession to the European Economic Community. It attempted to address the need for the mass production of housing by private capital. On the other hand, such measures hurt the small landowners and the people employed in the constuction industry and it was heavily criticized by specialists and by local authorities.

The period after 1981 saw new perspectives emerge. The 1981 elections brought into power the Panhellenic Socialist Movement (PASOK), a political party representing social modernization with the incorporation of some socialist elements. PASOK introduced a series of reforms in planning practice, and local participation in physical planning was extended both downwards and upwards (Lagopoulos, 1992). Downwards, new units and agents were created which participated in an advisory role in the decision-making of local authorities (for example, the neighbourhood chairperson), while simultaneously mechanisms were established for the activation and participation of the citizens. Upwards, a second level of local government, the prefecture council, was created and given important responsibilities in matters of urban, environmental and development planning within the prefecture (e.g. annual local development planning).

However, some experts have criticized the larger autonomy thus granted to the two levels of local government, on the grounds that it ignored the modern European needs of regionalization, in a state still largely protectionist which has not yet instituted essential reforms (Papadopoulou, 1995). There was still no institutional framework to link regional development in any essential way to regional self-government. Perhaps for that reason, there was little popular developmental consciousness or sense of local community, to counterbalance the traditional individualism of the citizens.

The result was that the principles of regionalization, which reflect the structure of the economy, were still almost the same as the ones existing before accession to the European Economic Community. At the same time, inflexible bureaucracy and the regional polarization still existed, although to a lesser extent, while the positive regulation associated with decentralization (e.g. unification of local communal authorities, development associations, etc.) lost much of their effectiveness on the level of practical realization.

The basic urban planning legislation (Law 1337/1983) passed by PASOK provided mainly a veneer of democratic planning principles, while there was still no national regional development plan (Koutsakos, 1993). This law was heavily criticized by various relevant organizations (TEE, 1993). Today a new housing law is being debated.

The basic urban planning law (1337/1983) established a system of General Urban Plans. Such plans are prepared on the initiative of local authorities and are concerned with the development of a particular settlement, and are supplemented by an Urban Study, which provides more specific plans for aspects of the General Urban Plan within the general framework established by the latter. These two plans are the tools used by administrative agencies and city planners to assess the needs of the 'neighbourhoods' (the units into which the settlement is subdivided) for public spaces, social services and public housing.

The passing of this law coincided with a large-scale initiative on the part of the Ministry of Regional Development, Settlements and the Environment, labelled Operation Urban Reconstruction, which involved the production of a large number of urban plans for settlements which either had never had a plan, or for settlements with plans that were woefully out of date and inadequate to cope with the, at times, explosive urban growth which occurred during the mid-twentieth century. However, Operation Urban Reconstruction was never given sufficient mechanisms of implementation and support (e.g. financial and technical resources, appropriate administrative agencies and adequate institutional framework). In spite of these limitations, it produced an impressive amount of work, among other things defining the boundaries of 7,871 of the 9,987 settlements of the country, drawing up plans for 539 towns with up to 2,000 inhabitants, and regulating the incorporation within city limits of some 132,000 hectacres of planned or unplanned constructed housing.

In conclusion, Greece has for historical and socioeconomic reasons developed a highly centralized system of physical planning characterized by inefficiency and bureaucracy on the part of the central administration, by a lack of essential participation on the part of regional and local authorities, and by a general lack of co-ordination. However, several scholars are sceptical of the new model of balanced regional growth which the government is now trying to implement, arguing that it may well be no more than a political ideology (Getimis, 1985).

The political, social and economic development of the country in the next few years will determine what role the state will play in the economy in relation to private capital in the future. This relation, in turn, will define the new political orientation of the country: will Greece take on the character of a social democracy or that of a formally modernized capitalist state?

The case study: Archangelos

The last few years in Greece have witnessed the emergence of a new field of sponsorship. Some private firms are currently engaging themselves in 'brotherhood' with villages or small regions. Usually such a brotherhood means simple donations to the village, but it also may culminate in deliberately provoking and eventually participating in the realization of projects of regional development.

A private firm manufacturing ceramic tile in Thessaloniki, Philkeram-Johnson S.A., decided to offer assistance to a small, distant and more or less stagnating community in the border area of northwest Greece. The search for such a community ended with the village of Archangelos. The firm and the village, through its Local Council, entered into a quasi-official kind of brotherhood. In the first phase of this relationship the company offered donations of a more trivial kind, such as repairing the school building, but it soon became apparent that this was not of great help to the community. It seemed that what was really important was a more global or structural intervention in community life.

At this point the firm asked for advice from the Aristotle University of Thessaloniki. A consulting team was formed and began working in 1992, in association with the community. The work, in both planning-proposing and monitoring terms, is still in progress.

The paradigm of the mountainous border community

We might perceive Archangelos as a paradigm for many villages barely surviving between stagnation and, at times, absolute decline. There are many such villages, especially in the mountainous and border areas.

In the case of northern Greece, an extensive part of the region is largely unsuitable for competitive agricultural development; it is thinly populated, with very steep mountainous terrain and insufficient communications networks.

Northern Greece became part of the contemporary Greek State comparatively recently. As noted above, Greece emerged as an independent state in 1827, after some four hundred years of Turkish rule. Northern Greece was liberated only in the second decade of the current century (1912-13). World War II broke out some 27 years later, followed by the German occupation (1941-45) and the civil war (1945-49). The civil war was an exceptional burden for the area because the northern border of the Greek state was the border to the 'iron curtain' countries as well. The late sixties were characterized by economic migration. Many young people migrated to foreign countries, mainly to West Germany. Internal migration to the major cities, mainly Athens and Thessaloniki, was also heavy. The political situation remained uncertain and thoroughly tense. April 1967 was marked by a right-wing coup d'etat, establishing the military dictatorship which only collapsed in 1974.

The sixties and the early seventies complete this turbulent period. The result was a rearrangement of the human and economic geography of the country. The seventies, and even more so the eighties, witnessed an increase in agricultural revenue, partly due to the subsidies paid to the farmers by the Greek government and the EEC. This led to a revival of the agrarian communities, but affected mainly those communities that were relatively strong, competitive and functional. The benefits were less visible in the under-productive mountainous villages.

The village of Archangelos

The village of Archangelos is the only village of the Archangelos 'community' (in Greece community, κοινότητα is the term for local administrative areas with small population, that is, administrative units with up to 2,000 total population comprizing at least one village, irrespective of their area). Archangelos is located very close to the border with the Former Yugoslav Republic of Macedonia. To reach the nearest town one has to travel either 38 km (30 minutes by car, 50 min. by bus) to the south-west, or 35 km (50 min. by car, no bus line) to the south-east. The south-west destination is the town of Aridea, with a population of some 4,700, and in the south-east is Axioupolis, with a population of 3,400. The most important links are to Aridea. The next adjacent major destinations are Edessa, with a population of 17,000, and Thessaloniki, population of 700,000 (Figure 10.2).

Figure 10.2   Regional location of the village of Archangelos

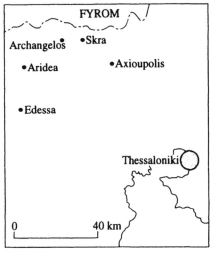

The community is situated at 820 m above sea level and the terrain is very uneven with steep slopes. The landscape is impressive with woods, creeks and a small river. It offers a characteristically wide variety of scenic views, even from inside the village. The village bears a long-time relationship with the nearby Monastery of the Archangel Michael, 2 km to the north-west.

Today, the population of the community is around 780; whilst there were some 870 inhabitants in 1951. If these data are accurate, it is possible to see a transition from decline to stagnation. It is also probable that there are signs of a recent slight recovery. Employment is principally in the primary sector, mainly farming. Sheep herding occurs in small numbers and only as complementary to other farming. There are a few small stores and recreational facilities in the village: a bakery, some cafes or restaurant-taverns, one or two places occasionally functioning as discos, and one or two corner shops. These retail activities are also complementary to farming. Two houses offer rooms to rent. These rooms, six in total, are used by hunters visiting the area in the hunting season.

Under normal circumstances, some 65 per cent of the total village revenue comes from selling cherries to fruit merchants, but this income is very unstable. The product is heavily weather-dependent and the absence of a producers' trading cooperative results in large price deviations and occasional difficulties in the marketing of the product.

The man-made environment is remarkable. In the village there are several old buildings, houses and accompanying structures (such as small stables, warehouses and the walls surrounding house compounds), with characteristic traditional architecture. There are even some coherent clusters of traditional dwellings. Given the relatively long history of the village, it comes as no surprise that the open spaces retain a typical traditional structure: a central square, the main roads running to the square and connecting the village with the major destinations, and a complex network of winding local streets. The church of the Monastery of the Archangel Michael was built in the 12th century and contains wall paintings of that era. The monastery, after eight centuries of active life, remained closed from 1912 until recently. Today there are two or three monks and a revival of the icon-painting tradition seems to be emerging. Having noted these features, it is important to note that neither the monastery nor the traditional buildings of the village attain major status in terms of archaeological interest or immediate tourist attraction, especially in view of the length of the history of Greece and the significance and fame of other archaeological sites and traditional villages.

As picturesque as is the physical environment of the village, both natural and man-made, a major problem remains. The network of open spaces was obviously functional in the pre-automobile era, but today the traffic and extensive car parking deprives it of its potential for social life. The main square, for example, frequently tends to deteriorate to something between a cross-roads and an informal parking lot.

The axes of intervention

The consulting team faced a problem that, in its essence, called for the definition of a series of actions that could help in provoking and sustaining community development. Some small-scale assistance, financial and other, would be provided by the firm of Philkeram-Johnson, but emphasis was placed on ensuring the self-sustainability of development. The problem has two main components: The economic base and the social life of the community. Securing and upgrading the community revenue, and improving the potential for social life would appear to be the necessary conditions for the long term survival and well-being of the community.

The economic base of the community. To secure and expand the economic base of the community the team focused on (a) the primary sector and (b) the tourist-attraction potential of the village.

a)   The risks and difficulties in both producing and marketing the community's main product, cherries have already been noted. However, an analysis of the primary sector potential of the community showed clearly that cherries are the most profitable possible product, given present international trading circumstances (namely the GATT agreement). To fulfil the potential of the community in this sector, some crucial factors must be taken care of. First, the poorly maintained elementary irrigation system of the area must be expanded, improved and refined. The cost of this action is relatively small, so the whole issue becomes a matter of exercizing political pressure on the higher-order local and central government agencies who are involved in the funding and construction stages. The irrigation system will drastically reduce the weather-dependency of production. Second, some action must be taken in relation to marketing. At the very least, this means adopting a common price policy towards the merchants. At best, it requires active involvement in the distribution process to the greatest possible extent, including packaging and transportation to the major distribution centres. Last, and certainly not least, comes the introduction of the farmers, especially the younger ones, to contemporary techniques in both farming and marketing.

b)   The community had rather high expectations for tourist development, mainly through public investment in a medium-size hotel building. This expectation was unrealistic given the condition of the village, the size of the investment, and the competitive attraction of other centres. However, it seems reasonable to proceed with a price-conscious, piece-meal development of the factors that could drastically improve the image of the village in this field: improving the network of open spaces, renovating the more interesting traditional building groups and mildly marketing the village, that is its history, landscape etc., through publications and local events. The public relations department of the firm and, of course, the firm's willingness to present its own contribution, is of great help in this domain.

These cautious actions in regard to tourist development will also, in general, positively affect and expand community life. This is especially true in the case of the network of open spaces: a successful rearrangement will re-establish the traditional use of open spaces as social contact and resting places.

The social life of the community. The significance of open spaces for the everyday and social life of the community have been noted above. Bearing in mind that, although it is set in a mountainous location, Archangelos is still a Mediterranean village, it is important to stress the role of the square as the focus of community life. The central square is the contemporary equivalent of the ancient agora: the main meeting place, the gathering place for religious or national holidays, the focus of economic life and the periodic market, and the major recreation spot with cafes and taverns. Whenever weather permits the customers of the cafes and the taverns sit in front of the buildings, in the square. As mentioned before, the central square is at present degraded by its use as informal parking space and by the annoying heavy through-traffic of cars and tractors. Restructuring urban space is important from the perspective of both tourist development and the upgrading of social life.

The community offers very little in the way of recreational and cultural activities. Besides visiting the cafes or the taverns, playing pinball or going to the discos (when open), there are no other real possibilities in the village. There are only a 'social centre' and a 'youth centre'. The first is occasionally used for activities engaged in by women and children. The latter, which was once fully functioning, is now almost deserted and is rarely used as a school for classes in traditional dances. It would be useful to invest these facilities with fresh appeal and expand the potential for cultural and social life.

Specific plans, actions and procedures

The consulting team focused mainly on three specific areas of action: the primary sector production issue; the restructuring of urban space; and the architectural tradition issue.

Following an analysis of the primary sector production problems and potential, it was decided to appoint an agriculturist as consultant to the village. This was a low-cost action with possibly very significant outcomes. The sponsoring firm has covered the cost and the agriculturist is now offering his services to the community. The possibility of taking control over the marketing stage is under consideration and is the subject of discussion with the producers. Other related actions, such as securing the expansion and improvement of the irrigation system, are also under way.

In parallel, the restructing of urban space is also proceeding but unfortunately the hierarchial radial pattern of open spaces, laid out over the very steep terrain, contributes heavily to the traffic problems encountered in the central square and the main streets. A poor distribution of open spaces gives the impression of high density, although in reality urban density is very low. The traditional buildings are hardly detectable in the current image of the village, they are concealed by various 'modern' or heavily modified buildings.

The consulting team proposed a 'structural interventions plan' (Figure 10.3). The main element of the plan is a walkway designed for pedestrians. Cars are allowed in most parts of the walkway, but only to allow access to the dwellings and with special steps taken to retain the pedestrian character of the space. The pathway interconnects the focal points of the village, helps to project scenic views, and invites the visitor to progress with anticipation to various goals through changing spaces and views. The pathway also provides opportunities for functions such as impromptu and more formal playgrounds, together with resting places suitable for chatting and social contact.

Figure 10.3 Structural interventions plan of Archangelos

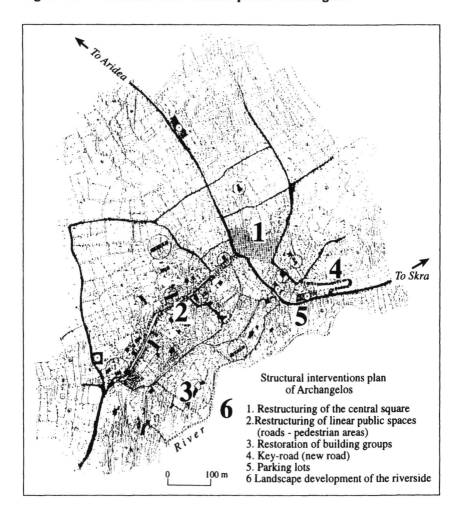

Structural interventions plan
of Archangelos

1. Restructuring of the central square
2. Restructuring of linear public spaces
   (roads - pedestrian areas)
3. Restoration of building groups
4. Key-road (new road)
5. Parking lots
6 Landscape development of the riverside

142

The central square, which is devoted to pedestrians, retains its reference status. Starting from the central square the pathway progresses south-west, along one of the main roads, to a new and currently dysfunctional 'square' (really just a newly paved open space). The renovation scheme incorporates most of the existing square plus the school yard, the churchyard and smaller lots of other publicly owned facilities. This new square, also drastically redesigned, offers an excellent view of, and the possibility of access to, the nearby small river and the traditional dwelling clusters on the riverside. Visiting the riverside is one of the two alternative branches of the pathway that the visitor can follow from the new square; both lead back to the central square. The system of paths covers most of the village and offers a range of opportunities to the residential areas that it passes through.

Most vehicle traffic is to be diverted by means of ring roads partly bypassing the central square. The purpose of this scheme is to accommodate the local traffic, that is traffic within the village and the community farms. It is intended that the main through traffic should continue to use the main roads to the east and west joining at the square, but it is important to note that local traffic is by far the major part of the total traffic load.

In order for these bypasses to fully function, both the 'key-road' and one of the two roads in the north-west quadrant of the bypass scheme must be functional. The key road is to be introduced by a structural interventions plan, while the other two roads are not yet fully realized. The two roads are part of the town plan prepared some ten years ago, but unfortunately this plan has not been officially accepted because of local reaction.

The structural interventions plan was presented in public and was officially accepted by the Community Council. The next step is to submit it to the higher order local government in a more formal version, as a 'local lay-out plan' (that is, a lay-out plan for part of a town). When this plan is accepted, after some time-consuming feedback from the community and from any citizen possibly affected by the plan, it will be possible to proceed to the construction stage. Currently the planning process is at the 'local lay-out plan' preparation stage. The structural interventions plan addresses the parking problem as an integral part of the urban space issue. The consulting team addressed some urban infrastructure problems as well.

The final element is concerned with architectural tradition. Morphological analysis of the traditional dwellings showed that they follow two basic rectangular patterns. These patterns, of both one- and two-storey, are variations of a major 'wide-facade' rectangular pattern. A major element of the facade and the entire composition is the porch, two-storey in the case of some two-storey dwellings, called hayati. As usual in the Balkans and Asia Minor, it faces south (at times south-west or south-east).

It is possible to restore some of the traditional dwellings. Some of them can function in ways appropriate for the tourist-development scenario, for example, they can be used as small museums, pensions, and show-rooms for traditional products or hand-made articles (such as embroidery and woven clothing).

In the first part of the pathway, the buildings on both sides of the street form almost continuous facades, as in a medieval street model, and in this case it would be both important and sufficient to restore or improve just the facade. The advisory team proposed such a plan, which will also allow for a wide range of variations for the development of each dwelling.

In addition to the above actions, a new 'local building regulation' was introduced. Compliance with this regulation will hopefully allow for new buildings to be constructed which are well suited to the local environment.

As far as housing is concerned, an unexpected issue arose: The village people consider the cost of a building permit as almost prohibitive and this acts as a disincentive to certain of new households in the village. This is a problem that can be met only at the central government level, perhaps as a form of subsidy to declining border areas.

Conclusion

One can view the case of Archangelos in a number of ways. Firstly, as a paradigm of the problems faced by remote mountainous communities (provided that Archangelos is not among the worse cases of the almost- or already-deserted villages) and as a case study on the conservation-modernization compatibility issue. Secondly, and in more abstract terms, the consulting team encountered several interrelated issues that are inherent in the process of planning. The Archangelos case clearly demonstrated the need for (a new) comprehensiveness, that is a multidimensional, or interdisciplinary, and which represents a global approach to community problems. It also made obvious the need for the construction of open-scenarios in order to encourage a flexible approach to planning. The importance of a rapid decision-making and implementation mechanism, capable of overcoming time-consuming bureaucratic procedures, was also apparent. Marketing the place or the project, in both the narrow and wider senses, also emerges as a crucial factor.

Finally, another issue, which is important both in general and specifically in the Archangelos case, is what we might call 'the leading-part question': Who induces, provokes or co-ordinates the development or development procedures? In the case of Archangelos it was a private firm which, however good-willed and helpful, is still an external agent. So, who will play a similar role in other communities? This connects to the issues of motivation or 'motivating', that is, in an era of decentralization and the withdrawal of central state initiatives, who will take responsibility for the development process and why, or how, can the motivation be provided which will induce such behaviour? Issues such as participation and conflict-management are additional aspects of the same problem.

Acknowledgement

The authors would like to express our thanks to the Planning and Environment Research Group of the Royal Geographical Society and to the Institute of British Geographers for their initiative in organizing the conference at which an earlier version of this chapter was presented. We should also like to express our thanks to Dr. Philip Kivell, who helped to make possible our presence there.

References

Benas, D. (1978, 2ⁿᵈ ed.), *Η εισβολή του ξένου κεφαλαίου σιην Ελλάδα,* (The Foreign Capital Invasion in Greece), Papazisis: Athens.

Christophilopoulos, G. (1979), *Το Δίκαιον της Ενεργού Πολεοδομίας,* (The Active Town Planning Legislation), doctoral dissertation, Athens.

Fontaine, P. (1995 2ⁿᵈ ed.), *Δέκα μαθήματα για την Ευρώπη* (Ten Lessons for Europe), European Texts series, E U: Luxemburg.

Getimis, P. (1985), Ο Σχεδιασμός του Χώρου στις Νέες Συνθήκες Γεωγραφικής Κατανομής της Διοικητικής και Πολιτικής Εξουσίας στην Ελλάδα', (Spatial Planning under the New Conditions of Geographical Distribution of Administrative and Political Authority in Greece), *Τεχνικά Χρονικά* (Technika Chronika), 4-6/85, pp. 37-39.

Kafkoula, K. (1990), 'Ο Αλέξανδρος Παπαναστασίου και η Ανοικοδόμηση της Ανατολικής Μακεδονίας , μια Προσπάθεια για τη Δημιουργία Προτύπων Οικισμών', (Alexandros Papanastasiou and the Reconstruction of Eastern Macedonia. An Attempt in Creating Model Settlements), in *Dec. 1986 Conference on Alexandros Papanastasiou: His Social, Economic and Political Views* (in Greek), Conference Proceedings: Athens.

Kafkoula, K., Papamihos, N., Hastaoglou, V. (1990), 'Σχέδια Πόλεων στην Ελλάδα του 19ου αιώνα' (City Plans in 19ᵗʰ Century Greece), partition of the *Scientific Yearbook of the Architectural Department of the Aristotle University of Thessaloniki*, vol. IB'.

Kalogirou, N. (1983), 'Η Ανοικοδόμηση της Θεσσαλονίκης από τον Ernest Hebrard, μια Επέμβαση στον Αστικό Χώρο και την Αρχιτεκτονική της Πόλης ' (Reconstruction of Thessaloniki by Ernest Hebrard. An Intervention in Urban Space and the City Architecture), in *Neoclassical City and*

Architecture (in Greek), 2-4 Dec. 1983 Conference Proceedings, pp. 245-52, Aristotle University of Thessaloniki.

KEPE (ΚΕΠΕ) (1976), *Πρόγραμμα Αναπτύξεως 1976-80, 11. Πολεοδομική Οργάνωση*, (1976-80 Development Plan. 11. City Planning), research group report, KEPE (Planning and Economic Research Center): Athens.

Koutsakos, B. (1993), 'Τα Γενικά Πολεοδομικά Σχέδια' (The General Urban Plans), *Τεχνικά Χρονικά* (Technika Chronika), 5/93, pp. 54-64.

Lagopoulos, A. - Ph. (1992), *Το Σύστημα Προγραμματισμού - Σχεδιασμού στην Ελλάδα*, (The Planning and Programming System in Greece), student textbook (mimeo), Aristotle University of Thessaloniki.

Lavedan, P. (1921), 'La Reconstruction de Salonique', *Cazette de Beaux Arts*, Sept./Oct. 1921, p. 245.

Papadopoulou-Simeonidou, P. (1995), 'Η Εναρμόνιση της Αναπτυξιακής Συνείδησης στην Ευρώπη', (The Harmonization of the Perception of Development in Europe), *Scientific Yearbook of the Architectural Department of the Aristotle University of Thessaloniki*, vol. ΙΓ', pp. 321-32.

Svoronos, N. (1964), *Histoire de la Grece Moderne* ('Que Sais-je?', no. 578), Presses Universitaires de France: Paris.

TEE (Technical Chamber of Greece) (1978), *Πλαίσιο Διαδικασίας και Οργάνωσης για τον Χωροταξικό Σχεδιασμό σε Εθνικό Επίπεδο*, (Procedural and Administrative Framework for the Regional Planning on National Level), study group report, TEE: Athens.

TEE (Technical Chamber of Greece) (1979), *Οι κατασκευές στην Ελλάδα*, (Building in Greece), June 1979 TEE Congress Proceedings, Vol. 2, TEE: Athens.

TEE (Technical Chamber of Greece) (1993), *ΕΠΑ. 10 χρόνια μετά Πραγματικότητες και προοπτικές*, (Urban Redevelopment Operation. Ten Years Later. Realities and Perspectives), June 27-29, 1993, TEE Convention Proceedings, Vol. 5, TEE: Athens.

11 Impediments to sustainable development in the environmental policy of East-central Europe: the example of Hungary

István Fodor

Introduction

The serious economic and social crises in East-Central Europe are a major factor impeding the realization of an effective and successful environmental policy. Sometimes this difficulty is exacerbated by political situations such as the ongoing Bosnian crisis in the Balkan Region. In this part of Europe, environmental crisis is closely related to the crisis of the economy.

In such a situation it is important to ask the question: can sustainable development be realized in the countries of East-Central Europe at all? This question can be answered in the affirmative, although it is difficult to provide a single way of achieving sustainable development in any of the countries. In theory economic restructuring and privatization can assist in the enforcement of sustainable development in environmental policy, however, practical experience demonstrates the opposite.

What did the political-economic systemic change mean in East-Central Europe?

Before 1945 East-Central Europe was characterized by the existence of a market economy at different levels of development. Besides the well developed system of all elements of the classical market mechanism, the region had a traditional work culture. After 1945 this economic model was replaced by a centrally controlled command economy system which eliminated the classical elements of a market economy (together with the associated institutional system) in East-Central Europe and divided the traditional elements of this model from the work culture. The systemic change which occurred in 1989-1990 thus had to establish a new institutional system for a market economy (Figure 11.1). The creation of this market system, however, was and still is hindered by many difficulties in several countries.

Figure 11.1 The transition of Hungarian economic control in the 20th century

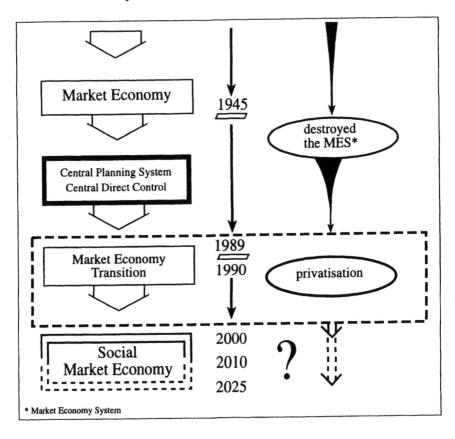

* Market Economy System

In addition, the countries of East-Central Europe inherited a strongly degraded environment, the state of which deteriorated after World War II (1945). The rate of and extent of deterioration was far greater than it should have been. This was a consequence of the centrally planned economic system in which the environment fell victim to politics and ideology. The degradation of the natural environment and natural resources was especially serious, mainly before 1980, because environment was considered a "free property" and a sacrifice to the needs of the economy.

Even though there has been a marked increase in social awareness about the quality of the environment following the political-economic systemic change, the restructuring of the economy, which started in 1990 and will be a long process, is likely to influence strongly, or determine the direction and intensity of changes in the quality of environment in East-Central Europe and Hungary. Although many ecological and restructuring problems have appeared in regions

Figure 11.2   Environmental problem areas in East Central Europe

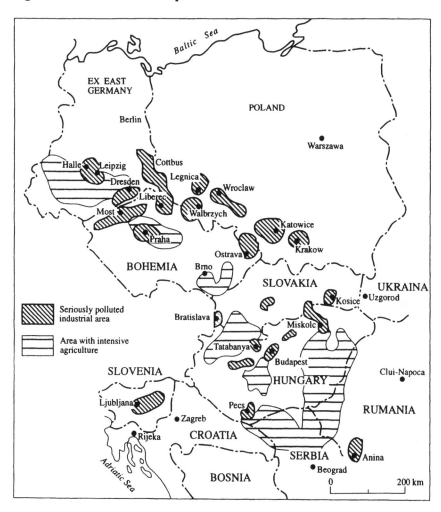

Source: Bassa, 1993.

which can easily be delimited - in most cases there is a direct coincidence between restructuring and the resolution of environmental problems (Figure 11.2) - an analysis of recent experience shows that restructuring which incorporates so-called environment friendly technology has slowed down and is now retarded or hindered by many factors. During the past few years, looking at the economic and ecological effects of policy, there are many examples of missed opportunity.

Following the political-economic changes, a process of democratization started in all countries of East-Central Europe; but this process occurred at varying rates and at different intensities. In this period, however, a number of new conflicts were about to emerge, for example, the rapidly growing level of unemployment (Figure 11.3) which had formerly been handled as a political factor and disguised within the overall functioning of the economy. The management of environmental policy was made even more complicated by a

Figure 11.3   Unemployment rates in East Central European countries (1989-1993)

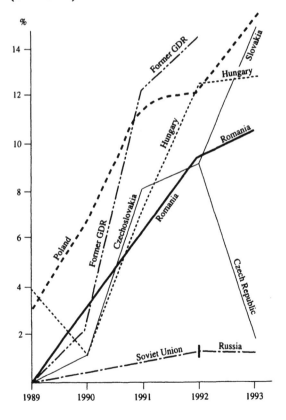

number of socio-economic difficulties which established a totally different background for environmental policy compared to that in the countries of Western Europe. In Western countries, the environmental and ecological crisis has been addressed in a stable economic-political environment. In contrast, the situation in East-Central Europe can be characterized by the following:

- environmental crisis management is attempted during a state of socio-economic transition.
- an unstable political, social and economic background.
- a low GDP per capita ranging from $2,000 - 7,000, the lowest level is experienced in Albania and the highest in Slovenia.
- additional problems are associated with political, social and economic restructuring including:
  - an economy in shambles;
  - a horribly outdated industrial infrastructure;
  - a critical housing shortage in most cities; and
  - a dreadful legacy of over-used resources and wide-spread environmental degradation.

This model of simultaneous economic, social, environmental and political restructuring in Hungary is depicted in Figure 11.4. The horizontal axis in Figure 11.4 shows time, while the vertical one indicates the environmental strain caused by different polluting sectors, using relative indices ranging from 0 to 1.0; the higher the score, the greater the emission of pollution and the more burdened the environment. It is almost impossible to precisely evaluate over the short-run the emergence and strength of the elements which have influenced the evolution of the quality of the environment. For example, agriculture only needed a few decades with its environmentally harmful technologies to pollute the first underground water layer (ground water). Whilst the polluting effects of chemical and other significant processes (large-scale animal breeding plants etc.) have decreased to a half or to one-tenth in some cases since 1990 and the rate of pollution of ground waters has slowed down, a solution to the problem of deterioration has not been identified.

In the field of industrial outputs affecting the environment, the situation is somewhat different. The decrease in industrial pollution, which experts estimate at 5-15 per cent in volume, is much less than that required by Hungarian environmental law and under the obligations declared in a series of international contracts. Looking at the evaluation of the environmental effects of the transition to a market economy, the negative aspects are strongly expressed in the literature whilst the only positive development related to the environment is the decrease in air pollution associated with bankrupt heavy industrial plants. However, even this can cause false optimism because, the

Figure 11.4 The trend of ecological crisis in relation to economic, social and political crises in Hungary

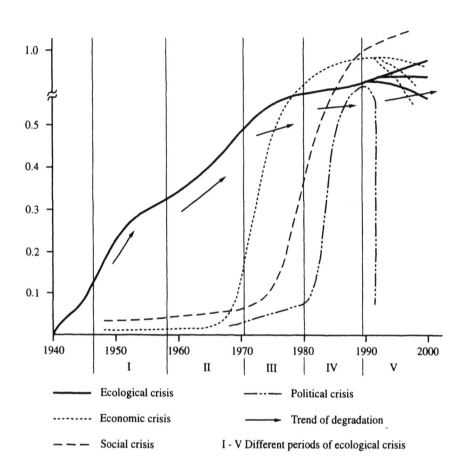

decline of air pollution is relatively insignificant compared to the dangers of derelict waste deposits. In the case of such waste no guarantees have been provided in relation to this huge problem as part of the liquidation and privatization procedures.

The standard evolution curves showing the political, economic and social crises can of course only be applied at a general level in East-Central Europe. In many countries political consolidation was very rapid, whilst in other countries, such as the states of the Balkan Region, it appears to be a long-term process.

Environmental effects of the changes in the ownership

The preceding analysis of environmental strains in the crisis regions of East-Central Europe has demonstrated the weakening of the business positions of the sectors and businesses which present the largest problems for the environment. Even if all other factors remained unchanged, the decline which took place in the production of hazardous emission has produced environmental benefits. Quite clearly this improvement is not the consequence of an enhanced environmental management system, but is a simple quantitative concomitant of a decline in production.

In contrast to this relative environmental improvement, there are negative phenomena connected with economic transition. Changes in both large-scale industry and agriculture have resulted in a mass reduction of productive jobs. In such circumstances, the importance of creating jobs has increased both from a business and an employment perspective. The influx of working capital favours job creation even if this results in the use of those technologies which generate undesired 'backyard' effects in certain regions.

Examples from Hungary prove that there were, and still are very few instances when critical cases can be effectively detected. The reason for this includes the fact that many privately initiated foreign trade activities, which result in job creation and the promise of capital inflow, are, in reality, a cover for business transactions that may be dangerous for the environment. What is a really unfortunate phenomenon is that society shows an exaggerated tolerance, which is unjustified even in difficult economic conditions, towards methods of job creation that should not be tolerated due to their negative environmental consequences.

The detection and elimination of environmentally adverse materials, technologies and consumer habits is extremely difficult in a situation where there are tens of thousands of business actors and hundreds of thousands of consumers. Research evidence suggests that the penetration of the private economy in East-Central Europe has not enhanced environmental awareness. In a situation of a strengthening private sector, an increase in state income was expected and this would have allowed environmental expenditure to be improved. However, an ever increasing budgetary deficit indicates that it is likely that the contribution of the business sector to the budget is decreasing, and this deficiency implies considerable disadvantage for environmental protection (which is still considered to be in a 'residual category', anyway). The amount of money received from the budget is clearly insufficient for systematic use and, as a consequence, resources are more suitable for financing only the most urgent projects.

The rapid growth in the number of businesses, privatization and the liberalization of business rules have increased the risk of the intensification of environmental stress. In such a situation, importance is attached to the collection and processing of environmental information and the development of monitoring systems. These additional systems require the provision of further capital assets only part of which can be covered by the support provided from the financial assistance provided by programmes such as the European Union's PHARE Initiative.

A number of examples in the fields of production and consumption demonstrate that the liberalization and deregulation of the economy have amplified the well-known 'backyard' effect or NIMBY syndrome. Included among the most evident examples are the large-scale import of materials hazardous to the environment, the transfer of technologies that are no longer permitted in economies where environmental regulations are stricter than in Hungary and the import of used cars in large quantities. This situation is made even more difficult by the fact that, within the developing market economy a number of supplementary monitoring and control functions have emerged, but the resources for the financing of these new functions is insufficient.

A fundamental and frequent problem is the fact that information is not always totally reliable. Many activities that contravene environmental rules are often concealed and therefore there is a great danger that, besides the boom in the number of businesses, the volume of hidden environmental damage is increasing. Even if such damage is discovered later, the assignment of responsibility and redress is difficult because of the possible liquidation of a company or due to identification problems. In all cases concerning the inflow of working capital, the import of consumer commodities and in industrial investment projects, if there is the slightest suspicion of environmental stress then the transaction is subject to legal control with respect to its environmental aspects. The Hungarian legal framework for this form of regulation is compatible with the requirements of the EU.

The elimination of the legal uncertainties related to land and property ownership is of vital importance for the assignment of environmental responsibilities. In this respect, the ownership reform carried out in the first half of the 1990s has brought about several consequences for the environment. In those cases where the owner can be identified the problems are less significant than in larger corporations. In such corporations many divisions and separate companies have been created within a holding company that has large capital assets and within which public and private ownership exists. A major problem encountered in these companies of mixed ownership is that the partly privately owned 'inner' companies attempt to transfer the consequences of any environmental damage to the state-owned 'cover' company. In a situation of this nature it is very probable that neither side considers themselves as really responsible for environmental damage and for the payment of compensation for such damage. This is the worst possible background for the definition of responsibility for environmental damages.

Another unfavourable consequence of the growing role of the private economy is the emergence of growing consumption and a squandering attitude in business calculations. Many businesses seem to be unwilling either to increase productive investments or to enhance profitability. This attitude makes it clear that the protection of the bases of production and the preservation of the unspoilt state of the environment are not included amongst business priorities.

In this economy in transition there is a considerable degree of exploitation, much of which primarily endangers the environment. Freedom of enterprise has unfortunately, been accompanied by a relaxation of public ownership and professional control and it is possible to identify many signs of the abuse of this situation.

The environmental aspects of progress towards the European Union

Two broad sets of conclusions can be drawn from the proceeding analysis and assessment. The first of these concerns the nature and status of environmental issues in the transition of the economies of East-Central Europe, whilst the second set of conclusions is related to the Hungarian case. Each of these sets of conclusions is now considered in more detail.

In the countries of East-Central Europe a number of broad trends and characteristics can be identified which indicate the future progress of policy, these include:

- In some areas pollution has stopped or emission is continuously decreasing.
- The rate of environmental pollution has slowed considerably in certain areas (e. g. industrial air pollution).
- However, alongside these favourable phenomena, new and often very significant sources of conflict have appeared.
- The general picture in the field of environmental regulation is very uneven; many of the countries of East-Central Europe are aware of the need for considerable investment to be made in order to raise their protection levels to that of the European Union.
- In other countries the preparation of new environmental regulations is in a more advanced state.

In the Hungarian case there is an emerging model which will assist in overcoming the environmental crisis, the features of this model include:

- Changes in the economic model including technological innovation.

- The development of environmental education and awareness as an important means of environmental policy implementation.
- Modernization of legal regulations in order to provide a basis for the urgent development of institutions for environmental protection and nature conservation; financial and management guarantees must also be provided as part of this programme.
- New elements of environmental regulation and nature conservation must be implemented.
- New areas of competence are needed by local governments in relation to the environmental management and development of settlements.
- Revision and improvement of environmental monitoring systems, modern systems must be available for environmental protection and nature conservation and such systems should incorporate reliable and easily applicable environmental information and data bases.
- Strengthening international co-operation in the field of environmental policy with special regard to the European Union.

References

Bassa, L. (1993), 'Environmental Management in Eastern Central Europe', *Geographical Newsletter*, Vol.1-4, pp.59-66, Hungarian Academy of Sciences.

Environment and Development, 1992, (National Report to the United Nations Conference), Ministry of Environment and Regional Policy Budapest, December 1991.

Enyedi, G., August J. Gijswijt and Rhode, B. (eds) (1987), *Environmental Policies in East and West*, Taylor Graham: London, pp.1-401.

Fodor, I. (1993), 'The Economic Dilemmas of Ecological Crisis Management', in Z. Hajdú, (ed), *Hungary: Society, State, Economy and Regional Structure in Transition.* Centre for Regional Studies, Pécs, pp. 69-78.

Fodor, I. (1994), 'Environmental Policy in the South Transdanubian Region', in Z- Hajdú, Gy. Horváth, (eds), *European Challenges and Hungarian Responses in Regional Policy*, Pécs, Centre for Regional Studies, pp. 507-14.

Part Three
POLICY AND MANAGEMENT

12 Environmental assessment and decision making

Fergus Anckorn and Nick Coppin

Introduction

Chapter 8 of Agenda 21 formulated at the 1992 UN Rio Conference highlighted the need to integrate environment and development in political and economic decision making. This chapter examines the contribution Environmental Assessment (EA) makes to decision making in the United Kingdom. Such decision making for individual projects is undertaken in a broader context of planning for sustainable land use. The problems and shortcomings of the EA process are examined in light of the author's experience with mineral resource development. The mineral sector faces a limited choice of suitable sites with fierce competition for land use and usually stiff opposition from the public, local communities and existing industry who are concerned about environmental impacts.

What is environmental assessment?

In this paper, the term Environmental Assessment (EA) is largely confined to the process/procedure that a project goes through in its planning and development stages, involving the developer, local, regional and statutory authorities, other organizations and experts. The outcome of this process is the written document, the Environmental Statement (ES). EA may also be applied to plans, policies and programmes (often termed strategic EA), but, as demonstrated here, its application at this level is conspicuous by its absence.

In 1969 the Congress of the USA passed the National Environmental Policy Act, which required all Federal agencies to include an environmental impact statement with all reports or recommendations significantly affecting the quality of the human environment. Since then, EIA/EA has become widespread throughout the world, mainly as part of the process of project appraisal. In 1985 the Council of the European Community issued a Directive (85/337/EEC) on the assessment of the effects of certain public and private projects on the environment. This Directive was implemented in England and Wales by The Town and Country Planning (Assessment of Environmental Effects) Regulations 1988 (SI 1988 No. 1199). Guidance on the procedures and the type of projects requiring EA are given in DoE Circular 15/88 (DOE, 1988) as well as a number of other publications (DoE, 1989; 1995).

Most EC guidance on undertaking an EA is either very general or conceptual in scope, or deals only with procedures rather than techniques. The UK DoE's Good Practice Guide (1995) is the most detailed of any existing EC publication and goes a long way towards remedying this.

EA and the planning system

The UK planning system is largely administered by Local Authorities (LAs) under the guidance of central government. LAs create a planning framework for future development by devising Development Plans and also maintain Development Control for individual project proposals. How is EA used in decision making under planning legislation?

Development plans

The use of EA in the UK is entirely restricted to decision making for individual projects. A proposed EC Directive on the Environmental Assessment of Plans, Policies and Programmes was not advanced beyond a draft stage, although some member states are still keen for this proposal to be pursued. The restriction of EA to individual project level is illustrated by the DoE Circular accompanying the EA Regulations which defines EA as:

> ...essentially a technique for drawing together, in a systematic way, expert quantitative analysis and qualitative assessment of a project's environmental effects, and presenting the results in a way which enables the importance of the predicted effects, and the scope for modifying or mitigating them, to be properly evaluated by the relevant decision-making body before a decision is given. Environmental assessment techniques can help both developers and public authorities with environmental responsibilities to identify likely effects at an early stage, and thus improve the quality of both project planning and decision making (DoE, 1988, p.1).

While the process of adopting a Structure Plan or dependent Subject/Local Plan is transparent and open for comment by all concerned parties, no provision is made for systematically subjecting the plan to 'expert quantitative analysis and qualitative assessment' to quote from the preceding definition.

Central Government does of course provide Guidance Notes as a reference point for both local government and developers. With regard to mining proposals, the local planning document in question is normally the local Mineral Plan and sector-specific central government policy guidance is supplied in the form of Mineral Planning Guidance (MPG) notes. However, these plans and guidance are formulated as a result of a general consensus on how to achieve a balance between meeting environmental policy goals (e.g. sustainable mineral development) and providing for the needs of (*inter alia*) the mining industry and communities. This consensus is base don the experience of the authors of the MPG's and consultees, without benefit of an independent EA.

With regard to the coal mining industry, Minerals Planning Guidance Note 3 (DoE, 1994a) provides little more than a set of factors which local authorities and industry should take into account in making their respective planning and development decisions. It suggests that in applying the principles of sustainable development to coal extraction it is relevant that, whilst coal is a finite resource, mineral working is not a permanent use of land and sites can be restored to a beneficial use of value to the community once operations cease. Consideration should therefore be given to the duration of the intended development and the proposals for restoration and to the extent to which the proposal provides national, regional or local benefits to the community which outweigh the disturbance occasioned during the development.

It may be argued that as a result of neglecting EA at policy and plan level, each and every development proposal subject to EA procedures will have to examine and justify (to some extent at least) the environmental consequences of the LA's Structure Plan/Subject Plan/District Plan insofar as it relates to the project in question.

Development control

Applications to the LA for planning consent are required for certain categories of development, to be supported by EA. The UK EA Regulations, in line with the EC Directive, specify Schedule 1 projects for which an EA is obligatory (refineries, power stations, etc.) and Schedule 2 projects where EA is required only if significant environmental effects may be involved. Mining developments fall into the latter category. Most new underground mining proposals would need an EA while surface operations will be judged on matters such as scale, sensitivity of location and whether anything novel is being proposed.

Many mining companies now volunteer to carry out an EA without seeking a ruling from the LA, seeing EA as a recognized forum for consultation and a tool for decision making in the widest sense. EA provides a structured framework for the company itself to consider the full range of issues. The resulting Environmental Statement (ES) documents responses from important agencies, such as the Environment Agency and conservation bodies, and provides data which the LA may need to reach an appropriate decision.

A project EA should go through a number of stages. Some applicants tend to overlook the earlier stages. Very often there is a last minute rush to prepare an ES, without time to compile sufficient meaningful baseline data, let alone to undertake proper scoping and consideration of mitigation and alternatives. While authorities continue to accept ESs prepared on this basis, developers do not have to go to the trouble and expense of commissioning a proper EA (even if not to do so is false economy in the long run). The quality of an ES (and by implication the EA on which it is based) is not a material consideration in determining a planning application, so the incentive for developers to undertake a proper EA from the start is less.

Research in the UK has shown that the quality of Environmental Statements for development proposals varies considerably (Coles et al., 1992). Lee and Colley (1990) developed review criteria for environmental statements, which grades them into six categories:

A   generally well performed, no important tasks left incomplete
B   generally satisfactory and complete, only minor omissions and inadequacies
C   can be considered just satisfactory despite omissions and/or inadequacies
D   parts well attempted but must, as a whole, be considered just unsatisfactory because of omissions and/or inadequacies
E   not satisfactory, significant omissions or inadequacies
F   very unsatisfactory, important tasks poorly done or not attempted

Using these criteria Lee (1991) found that there was a gradual rise in the satisfactory categories (A-C) from 34 per cent in 1989 to 48 per cent in the second year and 60 per cent by 1991. The percentage graded poor (E and F) declined only slightly in the same period, from 26 per cent to 23 per cent. Quality appeared to be affected by the size of the project, with larger projects producing better quality ESs. The Institute of Environmental Assessment use a modified version of the Lee and Coley criteria. Coles et al. (1992) applied these to 20 ESs and found the following:

Grade:	A	B	C	D	E	F
	10%	30%	30%	20%	5%	5%

This indicates that quality is improving, with 70 per cent being satisfactory, but still a significant proportion being poor.

The environmental assessment process

One of the advantages of the EA being the responsibility of the developer is that it can be closely integrated with the project design process. In order to allow project decision making to benefit, the stages of EA should therefore ideally follow the planning and design stages, as illustrated in Table 12.1 below.

Table 12.1
The stages of EA and project design

Environmental assessment	Project design	Authorities
Scoping	Project defined; pre-feasibility studies	(Initial consultations and soundings)
Project alternatives Baseline studies Environmental impact assessment	Feasibility studies and initial design Detailed design Restoration plan	Consultations Is a formal EA necessary? Provide baseline information
Environmental statement	Prepare and submit planning application	Formal consultations Request further information Determination
Extend EIA on some issues	Appeal Clarify designs	Appeal heard Decision
Environmental management plan	Full detailed design and contract preparation	
Environmental audit	Construction Operation Decommissioning	Monitoring Enforcement

EA is essentially a four-stage process:

Identification → Prediction → Interpretation → Communication

There are a number of variations on how these stages are practised, and the procedures that are followed, but all good EAs tend to follow this pattern. Identification involves scoping and defining the potential impacts, which sets the scene for the rest of the EA; how well this stage is done is therefore crucial to the quality of the EA. Prediction and interpretation are the heart of the EA, and communication is the presentation of the outcome (as an ES) to the public and decision-making authorities.

Each EA is different and the level of detail required (and thus cost) on each top will vary. Clients also have different perceptions of how detailed/thorough an EA needs to be. The appropriate level of detail required for baseline, prediction methods and impact assessment could be considered at three levels:

Level 1	-	Environmental appraisal (low key projects, peripheral issues; scoping of Level 3)
Level 2	-	Environmental impact assessment (non-rigorous, simple EA, key issues)
Level 3	-	Environmental impact assessment (rigorous, comprehensive EA, important key issues)

There could be a mixture of these levels in any one EA, dealing with different issues at an appropriate level of detail.

Project alternatives

There is an emphasis in the EA requirements to evaluate alternatives and to justify that the project as presented is the optimum environmental as well as economic option. This means doing at least a preliminary EA or scoping for several project options:

- Alternatives to the project *per se*, i.e. is it the best way to achieve the required outcome?

- Alternative locations for the project site.

- Alternative locations for the project components within the site (often constrained by the availability of land to the project).

- Alternative processes, waste disposal, transport arrangements, etc.

164

- The zero option, i.e. not proceeding with the project but continuing with the present situation (but considering any natural changes or unrelated actions that may be likely).

The first one or two of these are probably most appropriate to major or contentious projects. However, all EAs should consider the other three groups of possibilities. To what extent can a developer consider alternatives to/within his project? Is he willing to do so? There are usually constraints on alternatives due to land availability/ownership, cost, or the developer's proprietary process. Also, in the UK strategic planning will often have zoned land within a local plan, defining suitable locations for certain types of development (without benefit of the EA process).

Mineral deposits pose a problem in these regards, in occurring only at certain locations where working is possible, due to (*inter alia*) geological, hydrological/hydrogeological, engineering, and economic considerations.

Scoping

Scoping is one of the most crucial aspects of EA, yet it is one of the most neglected. Many inadequate EAs can be traced back to poor scoping. The objecting of scoping is to identify the potential impacts that are likely to be relevant (particularly key issues) and determine the appropriate level of detail for the baseline and impact assessment stages. It is also an opportunity to highlight potential problems and possible solutions (mitigation) early in the project's development. Scoping might be applied to a single project alternative, or might be applied to a number of alternatives and options.

It is surprising how few projects go through a proper scoping stage. Often consultants are asked to bid for an EA project where no scoping has been done; in which case they effectively have to go through a simple scoping exercise as part of preparing a bid, in order to arrive at a scope of work and therefore price. This is not very satisfactory, and perhaps the successful bidder is the one who defines the least scope of work, rather than the one who will prepare the ES to the appropriate level in the most cost-effective manner.

There is no set methodology for undertaking a scoping exercise. However, it will usually follow a similar, but much simplified, process as a full EA. At this stage it is important to clearly identify and define the impacts that are relevant to the decision making process. This is mainly an intuitive process, based on a checklist of potential impacts that has evolved from experience and consensus between authorities and environmental assessors.

Environmental sensitivity analysis

Environmental sensitivity analysis is a useful tool throughout the EA process, but particularly during scoping. This is a technique of zoning land according to its sensitivity to impacts arising from different sources and project elements. An example of this at the national scale is the groundwater vulnerability mapping being undertaken by the Environment Agency. This can also be applied at a local scale, by zoning the surface according to the risk of potentially contaminative releases from a project finding their way into groundwater resources. There are therefore two elements in the process: the relative vulnerability of the resource, and its exposure (in this example due to hydraulic connection) at the surface.

Zoning of this sort is best applied as a simple scale, such as:

Red - high potential impact and constraint; avoid, or only use with a high level of mitigation or if there is absolutely no alternative

Amber - moderate impact and constraint; proceed with caution and with mitigation or protection

Green - low potential impact, least sensitive and fewer constraints

An example of the criteria used to define these zones in the case of land used for tipping, , based on constraints on land take (due to human habitation, landscape quality, agriculture, ecology, archaeology) is given in Table 12.2 below.

Baseline - existing environment

The baseline defines the existing environment, against which any future changes will be measured. At the EA stage these changes are predicted rather than real. However, once a project is underway, the changes will actually be happening and the baseline will be an important starting point for measuring the significance of the project's contribution to any environmental degradation (or enhancement) that occurs.

The environment is variable in time and space, so the selection of the right baseline data to collect can be difficult. Some aspects of the environment can be characterized by one-off data, but most have seasonal characteristics or yearly variations that have to be included. There are two important criteria, or tests, to bear in mind when selecting data (Munn, 1979):

Relevance Only data that are necessary to assess the environmental impacts are necessary; an understanding of the nature of the potential impacts and the relevant environmental indicators is important; irrelevant data can hinder or cloud an assessment.

Table 12.2
Example matrix of environmental constraints
on availability of land for tipping

Environmental parameter*	Category of constraint		
	RED - over-riding constraints	AMBER - significant constraints	GREEN - low constraints
Humans (settlements)	Within 50 m of properties, particularly isolated dwellings. Intrusion on village identity.	Standoff from settlements and isolated properties: 50-250 m	Remote, at least 250m from properties. Visually enclosed land.
Landscape quality	Areas of high scenic value, including river valleys.	Integral units including farm fields and margins. Degraded river valleys and woodland	Derelict or disturbed land
Agriculture	High quality agricultural land (ALC1 or 2)	ALC grade 3a or 3b; active farmsteads and smallholdings	ALC grades 4 and 5. Farmsteads abandoned or with low intensity use.
Ecology	SSSIs, SINCs (county sites)	Significant wildlife diversity, e.g. hedgerows and scrub/woodland	Low wildlife value
Archaeology	Sites of national importance	Sites of local or county importance Visual evidence of unrecorded remains	Common/degraded or no sites; no evidence of remains

* many areas already protected in County Development Plan, e.g. settlements and protected routes.
Source: Wardell Armstrong, 1993.

Efficiency   Relevant data that can be organized in a reasonable time and at acceptable cost are preferred; this is secondary to relevance, but is an important practical consideration.

Most baseline surveys are a compromise between these two requirements. Data can be of various types and can be obtained from a range of sources (direct measurement, published information, remotely sensed images). In the UK, as in most developed countries, there is usually a lot of published information available. However, in developing countries there can be very little, and the EA team have to rely almost totally on field or photo-based data ('published' information can include that in the public domain and also that held by authorities or research bodies which is accessible to the EA team).

Project description

As well as giving a general description and context of the project, the project description should identify all the project's components and actions that are likely to be environmentally signficant. These components should be broken down into four stages of the project: construction, operation, decommissioning and post-closure; an example is given in Table 12.3. The project description should describe in detail the project's releases (emissions, discharges), outputs and consumption associated with each of the project components. It is these which will result in the environmental impacts that are assessed later, rather than the project itself. These need to be described in terms of their characteristics, constituents, qualities, timing, behaviour over time, release characteristics, media, etc. In addition, releases/outputs/consumption should consider those associated with normal planned operation (bearing in mind possible variations in operating conditions and efficiency) and those associated with accidents, emergencies and exceptional situations (cf. risk assessment).

Mitigation

Mitigation is the result of the interaction between EA and design teams, and the project as presented in the ES is a mitigated project. It is this mitigated project that should therefore be assessed for its environmental impact, because this is the proposal for which permission/authorization is sought. Nevertheless, the ES needs to explain what the potential impacts were, how these were addressed in the project design and the ways in which the project was modified.

Prediction of effects

The project outputs, releases and consumption will have an effect on the existing

Table 12.3
Example of project components and actions

Underground mining (metals)	Open-pit mining and quarrying (non-metals)	Landfilling of non-inert wastes.
CONSTRUCTION		
Infrastructure - access roads, rail link, power, water, drainage. Shaft sinking and underground development. Tailings dam, other earth structures. Tailings pipelines. Screening mounds, etc.	Infrastructure - access road, rail link, power, drainage. Processing plant, stockyard, workshops, offices, car parks, weighbridge, etc. Silt lagoons, water treatment ponds, other earth structures. Initial soil stripping. Screening mounds, etc. Soil stripping and stockpiling.	Infrastructure - access road, rail link, power, drainage, foul sewer. Offices, workshops, weighbridge, etc. Repository - earthworks, lining, leachate management system. Leachate treatment system or sewer.
OPERATION		
Ore extraction and raising. Underground ventilation. Mine drainage pumping, dewatering. Ore stockpiling. Ore processing. Tailings transport, disposal and rehab. Waste rock transport, disposal and rehabilitation. Concentrate stockpiling and transport.	Overburden/interburden removal, transport and stockpiling. Mineral extraction and transport. Mineral handling, preparation or processing. Mineral transport off-site. Overburden transport and backfilling. Progressive restoration of soils.	Waste importation and handling. Waste tipping and compaction. Leachate generation and management. Gas generation and management. Progressive rehabilitation and capping of completed cells.
DECOMMISSIONING		
Demolition and removal of surface structures and infrastructure. Removal of u/g equipment, sealing of mine entries. Final rehabilitation of tailings dam. Final rehabilitation of waste rock dumps. Surface drainage systems. Cessation of mine pumping.	Backfilling of open-pit, replacement and rehabilitation of soils. Removal of buildings, infrastructure, plant, etc. Rehabilitation of spoil tips, lagoons, water treatment areas. Surface drainage systems.	Final rehabilitation and capping. Regrading of settled areas. Removal of buildings and facilities. Installation of post-closure monitoring. Final surface drainage. Retention of leachate and gas management systems.
POST-CLOSURE		
Re-establishment of groundwater regime. Aftercare of restored areas. Long term land use established.	Aftercare of restored areas. Long term land use established.	Aftercare of restored areas. Continued leachate and gas management. Maintenance of drainage systems. (Completion certificate).

These projects components are typical of most operations of this sort. The final list will be specific to each project, and will be more detailed in most cases.

169

environment, i.e. the natural resources of air, water, land, biosphere and human environment. In other words, there will be changes in their character, behaviour, properties, constituents and quantities. In effect, prediction should forecast the new state of the environment. The EA regulations require the Environmental Statement to give a description of the likely significant effects, direct and indirect, on the environment of the development. These impacts are most conveniently considered under the natural resources which will be affected/changed, as shown in Table 12.4.

As a basis for predicting effects, it is important to be clear about what effects are significant in terms of impact. Each resource component should therefore have a series of indicators defined, which can be related to appropriate assessment measures.

There are many methods used for predicting effects on these natural resources and components, from numerical modelling, networks, overlays, graphic techniques, expert intuition, to outright guesswork. However, they should all examine how the releases/outputs/consumption is affecting the environment (i.e. targets). Consider pathways and destinations of releases; dispersal, attenuation, etc. and how these are affected by interaction with other factors, climate, geology, etc. The predictions should also consider the likelihood of emergency or hazard situations, as in risk assessment.

Assessment of impacts

What are we assessing the impact on, i.e. what is the environment? The basis for this is the natural resource components described above, though in effect impacts are considered in terms of their significance with respect of living things, especially humans. This is either directly, through effects on individuals, communities, culture and material assets, or indirectly through the ability to utilise/exploit the other resources (for materials, food, recreation, quality of life etc.). Therefore, we do not consider the impact of a project on, say, water *per se*, but on water quality as reflected in its ability to support human life and the biosphere on which humans depend for other resources. There are therefore many interactions between the resource components. Recently, the value of biodiversity has been recognized in its own right, for genetic resources and the health/vigour of the planet as a whole, without necessarily any immediate tangible value to humans.

The impacts predicted will have consequences (significance) for a variety of human interests. How is the significance of these consequences assessed? They are often described in qualitative terms - 'not significant', 'low', 'moderate', 'minimal'; these terms are not very meaningful unless they are defined against some standard or expectation.

Table 12.4
Natural resource categories and components

Natural resource	Resource component
Atmosphere	Air quality Noise Electromagnetic radiation Climate
Water (hydrosphere)	Surface waters Hydrologic balance (flooding, drainage) Groundwaters Coastal and marine waters
Land (lithosphere)	Land use and capability Soil resources Landscape Vibration Mineral resources
Flora and fauna (biosphere)	Habitats and sensitive areas Species/organisms Biodiversity/genetic resources Fisheries
Human environment	Social infrastructure Employment, economics Physiological and psychological well being Cultural heritage Material assets Buildings, infrastructure

Nature/characteristics of impacts There are a number of ways in which impacts should be evaluated and characterized:

- Objects (targets) of impact; *who/what and how many*
- Characteristics of impact; *direct, indirect, reversible, irreversible, cumulative, synergistic*
- Geographical (territorial) limits of impact; *site, local, regional, national international*
- Dynamics of impact in time; *short-term, long-term, continuous, periodic/episodic, emergency*

171

- Probability (and circumstances) of impact; *operational, failure, accident (risk assessment)*

Magnitude and importance of impacts

For some impacts, there are standards or thresholds to use (e.g. 55 dB(A) noise limit for residential properties). These standards tend to be black/white, i.e. they do not indicate any level or seriousness of impact, they just have to be complied with. They also vary from emission standards, to site boundary limits, to Environmental Quality Objectives which are not related to a specific development.

It is therefore important that each EA uses an appropriate set of impact criteria, relevant to the situation. Loose terms such as 'not significant', 'low', 'minimal', are meaningless unless they are related to criteria, and their use usually means that the impact assessment is short on proper objective analysis. However, for most impact areas there are no established impact criteria; this is a topic requiring some research.

Application of EA

Coastal Superquarries

Coastal Superquarries have been examined in some detail at both a polity and project level, and the superquarry concept provides an excellent sample of the use and neglect of EA in decision making. A costal superquarry is generally defined as having the capability to produce at least five million tonnes of aggregate per year for at least 30 years.

With the probability that access to quarry rock situated close to the main markets (i.e. the urban conurbations) in UK would become ever more difficult the establishment of very large remote coastal quarries came under consideration in the 1970s (Verney, 1976). However because of the huge capital outlay involved, the only UK coastal superquarry established has been Foster Yeoman's Glensanda Quarry on Loch Linnhe on the west coast of Scotland. The site was given full planning consent in 1982 and commenced operations in 1988. It was not therefore subject to EA procedures.

Meanwhile, the problem of securing sufficient rock to meet the UK's projected future needs in the face of growing public concern over environmental impact has continued to challenge the policy makers and planners. In 1991 the DoE commissioned a study to fill the gap in information regarding the potential of coastal superquarries. The report published in 1992 concluded that establishment of Scottish coastal superquarries was technically feasible and commercially viable (DoE, 1992). Yet no analysis of environmental issues was made.

Superquarries were now clearly in the Government's mind as a promising future supply option. This can be demonstrated by specific reference to coastal superquarries in MPG 6 published in April 1994, which suggested that the Government believes that subject to tests of environmental acceptability an increasing level of supply can be obtained from coastal superquarries during the period covered by this Guidance Note (DoE, 1994b). NPPG 4, the Scottish National Planning Policy Guideline on Lane and Mineral Working, also published in 1994 indicated that a maximum of four coastal superquarries would be appropriate for Scotland up to the year 2009.

While noting that the Government had arrived at these conclusions without benefit of an EA study of coastal superquarries, let us examine how industry has fared with its superquarry proposals under the MPG/NPPG guidance.

In fact, very few companies in UK have the resources to establish such a large capital-intensive venture, involving as it does, establishment of a large quarry site with major items of plant as well as a ship loading facility (and at least an arrangement for unloading at a suitable port). Following Glensanda, only one other scheme has been put through the planning application mill. This is the Redland Aggregates proposal for Rodel on the Isle of Harris. Although quarrying was carried out in the Rodel area in the 1960s, the superquarry project's history dates back to the early 1980s when several Scottish coastal quarry sites were being examined by the industry. Rodel received outline planning consent in 1981. Redland Aggregates submitted their expanded application for a superquarry in March 1991. At this time, industry protagonists were casting Scottish coastal superquarries as catalysts for regional development benefiting local industries and crofting communities and the Highland Regional Council's policy was to encourage the investigation and development of further coastal superquarries where:

- export is by sea or rail only;
- environmental consequences can be minimized;
- local job opportunities can be enhanced;
- deposits have a reasonable life expectancy;
- formal environmental assessment is made;
- a financial bond is secured for site rehabilitation at the construction phase and at the end of operations.

The Redland application was accompanied by an Environmental Statement and was duly subject to scrutiny by the Western Isles Council. With delays caused by requests for additional information and supplementary reports, the Council resolved on 23 June 1993 to inform the Secretary of State for Scotland that it was minded to grant planning permission. In reaching this decision, the Council considered the issues of visual impact, noise, vibration, blasting, dust, light pollution and pollution of water courses, all amendable to condition setting. In addition, the problem of Sunday working (in a Sabbath observing community) and potential impact of ballast water dumping were of concern and able to be dealt with through draft Agreements. On economic issues, the impact on tourism was addressed and the council considered that the greatest likely effect of the development would be the quarry's contribution towards a sustainable island economy. The Council concluded that there was no such thing as the 'perfect' development proposal, and the Rodel proposal was no exception. Advantages lay in the economic aspects, whereas various environmental factors were ranged against this ... the Council viewed the balance as being in favour of the development given adequate planning and legal controls.

The Council's officers report cited several potential grounds for refusal of planning consent, including:

- the detriment to the landscape caused by the proposed development (this could be either a view based on the specific circumstances of this locality, or a general policy view that developments of this scale and nature should not take place within National Scenic Areas, or both);

- that because of the uncertainties over the ballast water issue, it would be prudent to err on the side of caution and cite possible adverse effects upon the marine environment;

- the (existing) failure to reach agreement on the noise issue, with the attendant implication that the levels envisaged by the applicant company would be excessive in relation to the present background noise levels.

This list of reservations is significant in that they are not specific to the Rodel site, but to the concept of remote coastal superquarries in general. The Rodel EA therefore seems to have been saddled with settling issues which more fairly belong to assessment of the policy of encouraging coastal superquarries as a supply option.

The fact that the concept of coastal superquarries had not been properly investigated or settled at a strategic level was underlined by the reaction of the Scottish Office Minister for Roads two days after the Council's decision. He argued for some change in the pattern of supply and sugggested that in this particular context superquarries are not the answer, since they simply redistribute the problem from one part of UK to another, or abroad.

It would appear that the council's declaration also commenced a process of polarization of views among the local communities and heightened the interest of environmental lobby groups. It was perhaps not surprising then that against this background, the Secretary of State announced his decision in January 1994 to call in the Rodel application and hold a Public Inquiry.

The Inquiry commenced in October 1994 and ended in June 1995 (becoming Scotland's longest Public Inquiry) and was described by Kevin Dunion, Director of Friends of the Earth Scotland as the first significant test of its kind on sustainable development in Scotland. Redland Aggregates would probably have been justified in feeling at this stage that not only had the goal posts been shifted but the referee was playing a different game!

Without examining the pros and cons of the great range of issues re-visited at the Public Inquiry, this case study is sufficient to illustrate the need for strategic EA applied to the provision of minerals. It would appear to be the case that every such major quarry application will have to address national mineral development issues which remain to be settled. Until the Rodel planning application, coastal superquarries appeared to be considered by government as a benign quarrying option. The EA process has brought to light many issues which are clearly of significant concern. Companies, LA's and NGO's are now faced with the costs of re-visiting the debates at each Public Inquiry. The outcome of the Rodel inquiry is awaited at time of writing.

Opencast coal mining

The opencast coal mining industry in UK has had a comparatively short and eventful history since the first big earth moving plant became available in the 1950s. The industry was dominated by British Coal up to privatization in 1994 and until 1986, the Minister of Energy retained the final decision on planning consent. Despite growing concern over the proliferation of opencast sites, the Government determined in 1983 that the level of opencast coal production should be left to market forces and British Coal should provide a case for justification on a site by site basis. This was backed up by Circular 3/84 to provide site selection guidelines. This Circular highlighted the requirement for British Coal to present a case for need of the coal in terms of its quality and the market.

In 1986, Development Control of opencast coal mining was transferred to LA's with the Secretary of State for the Environment as final arbiter. Publication of relevant Minerals Planning Guidance (MPG 3) followed in 1988 together with the introduction of EA regulations at a time when British Coal was significantly expanding its opencast operations to around 18 million tonnes of coal per year, i.e. 20 per cent of UK coal production.

MPG 3 was first published in June 1988, as a set of guidelines designed to maintain the supply of coal from opencast sources without undue environmental damage. A new version of MPG 3 was published in July 1994, adding weight to provisions for environmental protection in line with sustainability considerations. MPG 3 provides:

advice to mineral planning authorities and the coal industry on how to ensure that the development of coal resources and the disposal of colliery spoil can take place at the best balance of social, environmental and economic cost, whilst ensuring that extraction and disposal are consistent with the principles of sustainable development (DoE, 1994a).

In the last two years, the LA's have firmed up planning provision through the creation of minerals local plans within the framework provided by their structure plans and MPG3. MPG3 requires minerals local plans to set out criteria for assessing individual projects, viz:

- employment and economic effects;
- effects on human activity, culture and amenity;
- effects on the biological and physical environment;
- effects on mineral resources;
- efficient mineral working.

The minerals local plans should also identify areas where coal extraction is acceptable in principle, when it is unacceptable and where coal resources should be conserved for future working.

None of the above national policy guidance or local planning provision is drafted with the benefit of an accompanying EA process. MPG 3 is drafted in line with cornerstone policy (sustainable development), and experience with mining operations through the years, as considered by its authors and the various consultees from government, industry, NGO's and the public. The hierarchy of local plans in relation to mineral working is developed in light of national policy and consensus on meeting local objectives and addressing concerns.

At individual project level, opencast coal mining is a Schedule II development with EA required only if the project is judged likely to give rise to significant environmental effects. Nevertheless, most applications for planning consent are accompanied by an ES, or at least some type of environmental appraisal is volunteered. Attention has focussed mainly on nuisance impacts (noise, dust, vibration, traffic), but economic issues such as the market impact of cheap opencast coal on underground mined coal have been raised. The issue of potential impact of surface coal working on air quality and community health has also been raised in recent years, but all evidence appears to indicate greater significance attaching to pollution sources outside UK.

All of these matters would appear to be amenable to resolution on a site by site basis as is current practice. However, it is most often the case that the proposed site is located in an area already affected by previous and existing opencast workings. Sites are often extended in area to extract coal outside the original boundary. Old sites may even be reworked to take deeper coal in an enlarged pit.

This potential for successive and simultaneous working raises the problem of long term cumulative impact of opencast working, which clearly needs to be addressed at Development Plan level rather than on a case by case basis.

Issues such as nuisance noise, dust and vibration and site traffic become more significant if the effect is likely to be chronic and long-term. It seems unlikely that cumulative impact can be adequately dealt with on an individual site basis and it emphasises the value of having plans and policies which do provide a sound basis for decision making and have been subject to the scrutiny and test of environmental assessment. Cumulative impact is directed to be addressed in Mineral Local Plans by MPG 3 but at present, this must be done without benefit of the systematic gathering, analysis and assessment of relevant data, consultation, scoping and consideration of alternative approaches which the EA process is designed to provide.

Conclusions

The current application of EA in UK is limited to individual projects. In the mining and quarrying industry, EA has been widely adopted as a useful decision-making tool both by the regulators and the industry and is applied in a manner which is consistent with EC guidance.

The mining and quarrying industry has received close attention from government with the creation of plans and policies designed to meet the goal of fostering sustainable development. Government has however so far shown unwillingness to adopt EA as a decision making tool in designing and applying mining policy and plans, and seems to have adopted a trickle up approach whereby experience gained at the development control level is used progressively to shape and influence their decisions.

The most obvious effect of this approach is that policy will always tend to trail behind practice. Where government took the initiative and, for example, sought to encourage development of coastal superquarries, the recent Rodel Public Inquiry experience appears to indicate that many issues had not been properly addressed - issues which would have been dealt with if EA had been applied during the government's policy studies.

The high profile opencast coal industry also appears to be illustrative of developers being forced to attempt to address environmental issues in project Environmental Assessments which would be more effectively settled at a strategic level, most notably in the common situation where the cumulative impacts of successive developments is a concern. The same applies to other situations such as aggregate quarrying in National Parks.

References

Coles, T., Fuller, K. and Slater, M. (1992), *Practical Experience of Environmental Assessment in the UK*, Institute of Environmental Assessment: London.

Department of Environment (1988), *Circular 15/88, Environmental Assessment*, HMSO, London.

Department of Environment (1989), *Environmental Assessment - A Guide to the Procedures*, HMSO, London.

Department of Environment (1992), *Coastal Superquarries to Supply South-East England Aggregate Requirements*, HMSO: London.

Department of Environment (1994a), *Coal Mining and Colliery Spoil Disposal*, Mineral Planning Guidance, Note 3, HMSO: London.

Department of Environment (1994b), *Guidelines for Aggregate Provision in England*, Mineral Planning Guidance, Note 6, HMSO: London.

Department of Environment (1995), *Preparation of Environmental Statements for Planning Projects that Require Environmental Assessment - A Good Practice Guide*, HMSO: London.

Lee, N. (1991), 'Quality Control in Environmental Assessment', in *Proceedings of the IEA/IBC Conference on Advances in Environmental Impact Assessment*, October 1991, London.

Lee, N. and Colley, S. (1990), *Reviewing the Quality of Environmental Statements*, Occasional Paper 24, EIA Centre, Department of Planning and Landscape, University of Manchester: Manchester.

Munn, R.E. (1979), *Environmental Impact Assessment - Principles and Procedures*, SCOPE 5, Wiley: London.

Verney, R. (1976), 'Aggregates, The Way Ahead', *Report of the Advisory Committee on Aggregates*, Department of the Environment, HMSO, London.

Wardell Armstrong (1993), *Landscaping and Revegetation of China Clay Wastes - Main Report*, Department of the Environment: London.

13 Quality assurance for planning and environmental management: the case for re-regulation

Michael Clark

Introduction

Both town and country planning and environmental management defy simple performance criteria. As Greed points out, not letting something undesirable happen may be a long term goal and a major, if intangible, achievement (Greed, 1993, p.5). How well planning performs depends on what it is expected to do. Viewed as bureaucracy, the planning system and arrangements for environmental management and protection are just another part of the public service. How effectively do these meet externally set, political, objectives? Do they impose unnecessary or avoidable delays? What does intervention cost? An organizational perspective draws attention to procedures, and to individuals' and departments' abilities to meet set objectives with specified resources and powers. In other words, it is managerialist in the sense of not questioning the wider structures which set the agenda and determine the extent to which this can be achieved (Knox, 1982, p.166).

Performance following these perspectives is addressed in terms of process rather than outcomes, and, while such rigour may increase administrative efficiency it does this in ways which protect, or fail to question, the underlying policies and interests.

> ... planners long-standing predilection for directing attention away from the substantive outcomes which planning is meant to achieve and towards a focus on the planning process. The result is that planning is legitimised as a method of decision-making and as a way of doing things rather than on the basis of what benefits or outcomes it is producing (Houghton, 1996, p.9).

179

This criticism is at odds with Khakee's statement that 'Most of the research and published literature deals almost entirely with the effects of planned intervention', though the narrow, instrumental, type of planning which he describes is distinguished from 'planning processes, rather than the policies or results of planning' which 'are hard or impossible to evaluate' (Khakee, 1994, p.441).

If planning and environmental management are intended to facilitate appropriate development, and to protect assets (environmental, ecological, aesthetic, archaeological, economic, cultural), they should be judged by their consequences: the quality and appropriateness of development. This applies whether these objectives emphasise protection of established communities and conservation of the natural environment (for example, calls to restrict development to damaged land and other 'brownfield' sites: Green Party, 1996), or whether they seek the removal of arbitrary constraints to commercial investment. Similar criteria fit strategic decisions and plan making (Department of the Environment, 1994). Every case requires appraisal of conditions prior to development (the 'baseline'), and some procedure for anticipating and evaluating the consequences, or impact, of development. In the language of EIA, screening should establish whether intervention may be necessary, and the criteria for appraisal. Scoping should identify significant potential impacts or damage, and may permit protagonists to at least agree about their differences. Subsequent decisions should benefit from monitoring of earlier developments and refusals, and the effectiveness of policies and plans, and from mitigation measures intended to overcome any problems which they might cause. This process of appraisal should be set in a wider context which looks at the overall effects of intervention including the total value added and potential achieved, or wrongs righted and assets preserved, during the period in which those being appraised were responsible.

Unfortunately, much of this is wishful thinking. For a start failure to develop may be noticed, but is less likely to be regarded as a fault of the planning system, or of any particular plan. Similarly, a problematic situation which continues despite policies or plans which are intended to rectify it, is likely to be blamed, rather than these measures. We should be judging the totality of what has been passed on from previous decision makers, not just those parts that have required explicit decisions to build, invest, restore or demolish. That is not to say that decisions won't routinely incorporate appropriate intelligence, or that they do not reflect the balance of professional and political opinion at the time. But they are unlikely to treat the status quo with the same rigour as proposals for change, or to deal comprehensively with what might have been (Clark, 1988). Nor can they draw on the resources necessary to cope with major planning problems or opportunities.

Planning problems

The rather depressing perspective developed so far can be countered by evidence of increasing rigour and systematization in the planning system. Strategic environmental assessment is evident in the routine environmental appraisal of development plans. Many planning documents are couched in the language of sustainable development. The legacy of derelict sites and contaminated land is being addressed by controls on the creation of new eyesores and polluted land. And there is a reasonable prospect that the new Environment Agency will be able to go further than the National Rivers Authority in identifying and rectifying deposits that threaten to pollute water suppliers, or cause harm in other ways.

However, contemporary British planning decisions are rarely heroic. We have learnt to put up with urban areas that are ugly, dangerous and unjust, and with a market that rations better quality education, health care, crime prevention and environmental amenity on geographical lines, which increases social segregation and generates unnecessary traffic. We may not regard this as the self perpetuating, grinding inequality and injustice that Bunge describes as the cities of Superfluity, Need and Death

> The City of Death is filthy. Abandoned homes punctuate the landscape. Rubbish decorates the berm. ..... Chronically unemployed slum youth (50 per cent and more) populate porches and curbs.

> The City of death is going downhill. Its homes, its streets, it schools and its children's' teeth fall apart as the money is sucked out (Bunge, 1975, p.162).

But the image is too close for comfort to be complacent about the effects of half a century of planning, especially if one takes a global perspective. Homelessness remains a serious problem. Many people have little choice but to accept unattractive, unsuitable accommodation. Piecemeal attempts are made to ameliorate what are largely inner urban problems, and to devise development plans which might rectify them - given the right mix of commercial initiative and political will. But in the UK planners and politicians have been so badly stung by the failures of previous bold attempts, and so tightly constrained by limits on public sector spending, that the last quarter century has seen little more than a patchwork of incremental, poorly integrated, often underfunded and rarely community-based attempts to restore selected parts of our inner cities and urban areas. Nor have any new towns been designated, though similar development commissions have let urban development corporations over-ride local democratic control, resulting in spate of speculative 'docklands' and 'brownfield' redevelopments. Many of these have created, or threaten, new planning difficulties for adjacent areas.

Meanwhile, an under-regulated property market permits the worst of the built stock to slide into decay, and takes countryside for a market determined mix of development dominated by motorway junction retail and office complexes, extensive road based industrial and warehousing sites, and an overwhelming mass of 'executive' four bedroom (just about) detached houses. All this at a time when official policy has accepted (and academics have long argued) that more urban fringe development generates vehicle movements and car dependency. We are told to plan for less travel, and to manage demand (Department of the Environment & Department of Transport, 1994, 1.4 -1.9), yet Government has little real influence on key investments in retailing, or for that matter most other types of business.

Similar complaints can be made about policy for the countryside, though here it is perhaps more appropriate to focus on environmental policy failings. Lack of effective control on farming, and inadequate ecological or archaeological protection against commercial agriculture, has ravaged the British (and particularly the Southern English) landscape. Intensive, high input, industrialized farming, large scale forestry, fish farms, big quarries and tourist boards' promotion of mass tourism in the 'picturesque' countryside have done little to ameliorate a rural crisis of falling employment and unbalanced population. Various systems designate important ecological and heritage sites. But funds or powers for comprehensive or long term protection are inadequate, despite legal and planning measure to preserve scheduled monuments and listed buildings, or the requirement to respect conservation areas and to refer to the importance of Sites of Special Scientific Interest (SSSIs) when preparing development plans or making individual planning decisions (Department of the Environment, 1994, pp. 22-24).

Passive regulation versus positive planning

Overall this adds up to a problem. While the Planning and Environmental Management system appears well staffed, professional and competent, its outcomes are far from satisfactory. There is a gap between individual development plans, which are generally realistic, well informed and sensible in their allocations and priorities, and the day-to-day reality of the places which most of us use. What is most disturbing about this paradox is that the purposive, forward looking and essentially optimistic traditions of town planning have become incorporated within a system of bureaucracy and routine political decisions which is essentially reactive. Its mission is not to survey the world, identify its faults, find solutions and implement them. Rather it has become the regulation of narrowly defined types of incremental change.

The 1994 Planning Charter Standards state that the planning system:

is used to make sure that things get built in the right place, and to stop the wrong things getting built. It makes sure that new uses for land and buildings are right for the location. The system helps to plan for the developments the country needs - the new homes, factories, offices, roads and schools. At the same time, planning has to protect the natural and man-made environment. It helps to make sure that development and growth are "sustainable" - in other words, that planning decisions will not damage the environment for future generations (Department of the Environment, Welsh Office and National Planning Forum, 1994, p. 4).

It is doubtful if the system can deliver on 'big' issues such as sustainability, even if one accepts the Department of the Environment's preferred definition:

... it should ensure that the best of today's environment is not denied to future generations (PPG12, 1992, N3 para 1.8 in Owens, 1993, p. iv).

To have any chance of achieving sustainable development objectives planning must take an integrated approach to environmental management. This should include some modification to private property rights and liabilities, better use of taxation, subsidies and grants, more effective monitoring, with its results publicly available, and appropriate regulation (Hall, Hebbert and Lusser, 1993, p. 28).

Planning also appears flawed in its ability to cope with routine, day-to-day matters. Action within the planning system has little influence or control over the key decisions: public sector funding (which affects the quality of regulation and intervention as well as public service provision and investment); provision and operation of transport and infrastructure services; energy policy; and the neglect and demolition of property as part of a wider process of speculation in development land. Even the mix of house types constructed or the use of commercial buildings is difficult to control. The system relies on commercial initiative to achieve almost anything. Local authorities have become so strapped for investment capital that some rely on 'planning gain' to achieve even modest objectives, arguably beyond their legitimate powers to seek compensation for the identifiable costs of a particular scheme.

Implementation

How a political or ideological objective is to be put into effect raises questions about the choice of administrative instruments, their ability to deliver, the extent to which objectives are agreed or consistent, and whether implementation may be problematic (Greed, 1996; Walker and Pratts, 1994). There is a fundamental tension between market based, market serving systems of passive regulation and more 'positive' or interventionist forms of governmental control. Circumstantial evidence suggests that deregulation is often an excuse for less regulation: for reducing the protection given to public goods or interests on the assumption that market discipline will lead to greater efficiency, and that the safeguards provided by regulation are unnecessary, or have costs which exceed their benefits.

This approach is illustrated by Regulatory Impact Analysis, a requirement that projects be undertaken if and only if their benefits exceed their costs. This specification was written into the United States' Flood Control Act of 1936 and is now embodied in administrative policy in the form of Executive Order 12291. Section 3 of this order states in part:

> ...each agency shall, in connection with every major rule, prepare, and to the extent permitted by law consider, a Regulatory Impact Analysis... each ...Regulatory Impact Analysis shall contain the following information:
>
> (1) A description of the potential benefits of the rule, including any beneficial effects that cannot be quantified in monetary terms and the identification of those likely to receive the benefits;
> (2) A description of the potential costs of the rule, including any adverse effects that cannot be quantified in monetary terms, and the identification of those likely to bear the costs;
> (3) A determination of the potential net benefits of the rule, including an evaluation of the effects that cannot be quantified in monetary terms.

Section 2 of the order states:

> ...in promulgating new regulations, reviewing existing regulations, and developing legislative proposals concerning regulation, all agencies, to the extent permitted by law, shall adhere to the following requirements:
> (b) Regulatory action shall not be undertaken unless the potential benefits to society for the regulation outweigh the potential costs to society;
> (c) Regulatory objectives shall be chosen to maximise the net benefits to society. (Freeman, 1982, 49, pp. 51-2).

Data gathered as part of regulatory impact assessment will help to determine incremental benefits and costs, identify trade-offs at the margin, and expose opportunities for more cost-effective regulation.

The ideology which favours less government intervention and prefers market solutions has led to widespread privatization and deregulation. Whether motivated by hostility to public sector spending (especially the PSBR), an attack on 'outdated' management styles and working practices, or a wish to 'break' union power, deregulation has gone beyond critical review of regulatory impact. It assumes that commercial provision is superior, even when direct transfer of public sector monopolies to the private sector permits the new owners large 'windfall gains', and lets them engage in such undesirable activities as asset striping, concentrating on profitable services (and demanding subsidy for others) and predatory cross-subsidization.

Deregulation and re-regulation

If deregulation and its impacts lead to, or coincide with, falling service standards and other difficulties then re-regulation appears necessary. However, this is difficult to pursue in the absence of clear or agreed service criteria, and when many of the resources and powers necessary for service delivery have been lost during privatization. It may be incompatible with the dominant ideology, despite circumstantial evidence and popular hostility to corrupt, regressive and damaging change. Even if the case for re-regulation is accepted, reliance on the private sector for development initiatives, investment, management and 'customer level' delivery add to the difficulties faced by an essentially reactive planning system and public sector.

Given widely held (if misguided) belief in 'the 'market', and hostility to 'socialist intervention', re-regulation may only be feasible if introduced in a new way. Here it may be worth considering recent academic and political interest in regulation, and the links between some of these forms of intervention and the more individualistic forms of intervention associated with quality assurance (QA).

The regulation school

This is an approach to social and economic relationships which prescribes 'nothing less than an anti-Taylorist, post-Fordist revolution. Workers would be made formal participants in decision-making, their commitment to the system would be sought by enriched forms of work and guarantees of job security and welfare benefits' to counter 'the attempt by capital to resolve the crisis [of having exhausted its potential for growth, which Lipietz attributes to its abuse of labour and ecology' (1992, p. 17, p.53)] by establishing a system of 'global Fordism': exploitative, neo-Fordist strategies of globalized, decentralized production, flexible specialization and devolved management (Kumar, 1995, p. 56).

The crises of deteriorating public service provision, loss of public goods and worsening terms and conditions of employment associated with deregulation and privatization, and with other macro-economic effects of Structural Adjustment, readily fit the Global Fordism model, so do the loss of planning powers and

185

general decline in the effectiveness of the public sector. Anti-Taylorist, post-Fordist solutions which stress individual empowerment, community development and the micro-scale imply new ways of organising society, work and the economy as a whole. Whether essential changes in attitudes and behaviour, and in institutional arrangements, can be achieved by the people directly involved, 'bottom up', or whether they depend on macro or Global scale change, is crucial to the success of any form of flexible co-operation (Standing, 1992).

From a 'regulationist' perspective, overcoming international capital is difficult, and depends on fundamental changes in the 'mode of regulation' to permit organized labour re-establish itself through a new form of 'class compromise' or 'social contract'. Capitalism is characterized by 'constant flux' and tends to 'transform itself in such a way as to give itself more space and more time'. (Kumar, 1995, p. 57, p. 164). There may be alternatives to domination by global capital or organized, paid labour, but self-help, the family and the informal sector appear increasingly disadvantaged. Is it realistic to assume that community-based, grassroots or individual action can challenge the technological, cultural and institutional domination of big business? Alternatively, 'soft technologies' and human scale types of organization may offer viable alternatives to the current, wastefully polluting, exploitative and inflexible, systems of industrial production, distribution and social organization (Dauncey, 1988).

Quality assurance and total quality management

Quality Assurance is part of the language of modern management, so might be seen as part of the problem. Management serves the interests of capital, and has been a key player in implementing strategies of social control which have destroyed labour solidarity, taken away much of the autonomy, professionalism and craft skill associated with paid work, and worsened many people's conditions of work and job security (Hutton, 1995, p. 105; Lipietz, 1992, p. 35). They have favoured the 'machine' economy of global mass production over less complicated and more locally based rivals. In the process we have seen further environmental degradation, excessive resource use and localized unemployment amounting to social and environmental 'dumping' in response to market opportunities.

But the language of Quality Assurance is benign and its objectives are unexceptional. It is 'the process whereby customers, producers or any other interested parties are satisfied that standards will be consistently met' (Ellis, 1993, p. 6). Quality is 'fitness for purpose':

> The totality of features or characteristics of a product or service that bear on its ability to satisfy a given need (BSI, in Ellis, 1993, pp. 3-4).

Ellis suggests that all QA should:

1   specify standards

2   identify critical functions and procedures necessary to achieve these

3   rely on consumers to establish if standards are met

4   show what standards are to be achieved, and which procedures must be followed

5   adopt a cybernetic approach to setting standards and procedures, with negative feedback leading to appropriate, effective action

6   involve all personnel, and be committed to development and training.

(based on Ellis, 1993, pp. 7-8).

TQM (Total Quality Management) is '..an attempt, a strategy, to produce an institution-wide commitment to quality assurance' (Barnett, 1992, p.117). Individuals are encouraged to take responsibility for their own actions, and to work co-operatively in non-threatening situations where the onus is on identifying and rectifying mistakes. This ethos is quite different from more hierarchical, sanction based management systems, and it is bizarre that employers may seek to impose quality assurance in a attempt to increase workplace discipline and reduce employee autonomy or professionalism.

UK higher education is fertile ground for such criticism, which perhaps reflects the extent to which the jargon, rather than the spirit, of quality assurance has pervaded its administration. Barnett criticises the tendency for QA to impose inappropriate standardization and discipline (1992, Ch.7). QA may reinforce the tendency for individuals to have to justify any failings, and respond to criticism, without passing blame 'up the line' to those responsible for resource allocation, task setting or any other 'structural' considerations. In these circumstances it becomes difficult to criticise arrangements for measuring educational performance without engaging in institutional politics. Laurie Taylor provides a humorous critique that reflects the zeal of some proponents of QA, and which also illustrates its potential for abuse.

> ...a precondition for advancement in TQM is that we successfully diagnose the variations in the system and decide which are due to variations in human abilities and which are caused by the system..
> ...the only way to survive is through learning how better to manage

resources...

...we can only move ahead by successful problem identification, data gathering, data analysis and the generation of proposals for solution, implementation and test.

It is only by insisting on quality and continuously improving Quality that we can achieve true Quality in every area of Quality and continue the cycle of never-ending improvement of Quality. That couldn't be clearer could it, Dr Piercemuller?

(selective quotation from Taylor, 1994, p. 95, p. 127)

The tedium and perceived threat of 'quality' procedures in Higher Education combine with cynicism about its domination by non-academic agenda (efficiency, workplace discipline) to create an intellectual climate in which it is difficult to be positive about QA. This is unfortunate as it has considerable potential, if used correctly. Although David Bellamy's face and London Transport have appeared on recent covers of Quality World, as has the title BS7750 Time to Manage the Environment (Hodgskinson, 1995; Hughes, 1995), this quality professionals' journal still reflects engineering and management roots, and lags behind the environmental literature and academia in its awareness of environmental management systems (Hodgskinson, 1995; Bennett, 1995; Hughes, 1995; Netherwood, 1994, 1996; Willis, 1994, p. 14).

Knowles ilustrates the universal nature of Total Quality Assessment, arguing that quality professionals' work will never be done,

> ... as a growing world population demands higher living standards and a better quality of life, and human ingenuity continuously contrives more effective means to serve society. The attainment of quality presents a moving target, requiring an ongoing commitment to improving standards ...to develop and apply ever higher quality standards.

It is their responsibility to ensure 'all forms of waste are minimized', to conserve materials and energy, and to be 'leaders in the sociological and political endeavours to oppose the unnecessary scrapping of things, squandering of assets and wastage of resources'. Too much reliance on standards may inhibit innovation. A quality management system should be judged by its results:

> better, more competitive products and services, growth in market share and better returns on investment - in short the creation of more real wealth (Knowles, 1992, p. 64).

If such claims are correct, QA, TQM and related management tools are the latest innovations to achieve economic growth through improved technical performance, a 'costless' Pareto benefit, so any organization not incorporating such improvements represents a real cost because of the opportunities foregone. If tools are neutral, then the early adoption of QA helps explain the 'machine' economy's competitive advantage over more traditional types of social and economic organization (eg. non-monetary exchange, unpaid work in the family and community). Similarly, the decline in public service provision and extension of the private sector may be explained by the greater efficiency of commercial organizations, though this is a more awkward case to argue as the motives behind privatization appear dogmatic and self interested.

Monitoring performance

Performance criteria may appear as an assault on those parts of the public sector that remain, in effect an attempt to routinise, and so devalue, craft or professional skills. Published 'league tables' obscure important reasons for differences in performance (for instance access to resources, the nature of the problem and the suitability of the criteria used for assessment). They may deflect resources from the most needy (or highest value added) situations, and reinforce the position of an established elite. But in its favour, monitoring should ensure that stated objectives are met, and provide independent, unbiased information for policy development and appraisal.

It would be unwise to read too much into the 'scientific' nature of commissioned research, but the Department of the Environment's 1993-4 Planning Research programme indicated priorities that are relevant to this chapter because of the prominence it gave to attempts to measure and evaluate the performance of planning and environmental management. If this research 'delivers' we can expect policy applications in these areas. The programme supports Delafons' statement that policy initiatives or priorities are not set by officials. Since 1985 little environmental or planning policy has been generated by the Department of the Environment (Delafons, 1995, p. 88). Morphet agrees, but explains this by the leading role of the European Commission (Morphet, 1995, pp. 199-200).

In seeking to inform current Ministerial decisions, guide the execution of policy, monitor the achievement of environmental goals, address future issues and augment industrial research (Office of Science and Technology, 1992, in Delafons, 1995, p. 85) the Department's research rationale led to 15 new projects for 1993-4, including:

* monitoring the effects and effectiveness of environmental assessment, derelict land prevention, planning control enforcement and the 1992 Planning Directions;
* policy development for travel patterns and behaviour, London's economic role and the quality of its urban environment;

- improved spatial referencing standards for GIS;
- identification of UK Planning achievements and trends since 1976, and an evaluation of the land use planning research programme since 1989;

and, perhaps most important:

- Improving the effectiveness of existing policies.
- Good practice guide for the Application of Sustainable Development principles in the Planning System;
- Realism of Development Plan Policies;
- Evaluating the Effectiveness of Land Use Planning (I): Developing Performance Indicators and Other Measures to Evaluate the Effectiveness of Land Use Planning;
- Evaluating the Effectiveness of Land Use planning (II): The Trade-off Between Economic and Environmental Considerations in Planning Decisions.

(adapted from Department of the Environment, 1994 in Delafons, 1995, p.88)

Appraising professional competence

This drive for efficiency and effectivness is paralleled by initiatives such as the Construction Industry Standing Conference (CISC) Mapping Exercise. This is a wide-ranging application and testing of performance criteria which emerged from CISC's involvement in vocational training to apply to competence in an increasing variety of professional specialisms (Greed, 1996, pp. 298-302).

> Most built environment professionals did not seem too bothered about NVQ as long as it stayed down at the manual and trades levels, and did not affect the higher professionals.... Now like rising damp, it is making its way up to levels 4 and 5..... Originally it was imagined that level 4 would be the quasi-professional higher technician level and level 5 would be the chartered body level (Greed, 1996, p. 300).

The CISC 'list of functional elements and critique' intended to 'establish, maintain and modify the use of the natural and built environment, balancing the requirements of clients, users and the community' has some 75 elements, ranging from

- Formulate strategies for the environment
- Monitor and review environmental changes and needs
- Establish mechanisms for monitoring and reviewing changes and needs in the environment

to

- Investigate ecological and environmental factors relating to the natural and built environment
- Prepare and agree an investigation schedule

to, finally,

- preparing forms of contract, drawings and schedules, specifications and bills of quantity.

(adapted from Greed, 1996, Appendix 2).

Much here is reminiscent of early, catch all, attempts at EIA (eg. the Leopold matrix; Wood, 1995, p. 196), and other elements are similar to, though far more prescriptive than, environmental management systems such as BS7750 and EMAS (Clark, Burall and Roberts, 1993, p. 140; Netherwood, 1996). While the CISC specification outlined above goes from the general to the particular, it omits:

1   Evaluation of conditions prior to intervention, or critical input to policy objectives and priorities;

2   A cybernetic approach: positive and negative feedback mechanisms to ensure faults and sub-optimal situations are identified and acted on.

It would be wrong to criticise the construction and surveying professions for not acting outside their brief. But the CISC map does indicate that institutional and professional processes are more suitable for performance evaluation, and for use of QA type procedures, than their output or outcomes.

Minimum service criteria

These are a crucial part of the regulation of newly privatized utilities, and of the separation of 'poachers' from 'gamekeepers' in many services which were until recently both provided and controlled by public sector institutions. The basic idea is that the regulator establishes a minimum service level which any operator is obliged to meet. For transportation, criteria are likely to include safety considerations (eg. vehicle specifications) as well as service frequency, quality of service provision and perhaps cost. While advocates of privatization argue that competition will oblige operators to exceed these standards, market forces may have the opposite effect, leading to services at the minimum permitted level, or at risk of being withdrawn unless subsidy is provided from public funds.

In an ideal world, all public transport would run to time, service frequencies would make late running or missed buses and trains unimportant, and we would wait in attractive bus, tram or train stations, and at very least would have the convenience of a shelter at our neighbourhood bus stop. Reality is different. Commercial and political considerations mean that many services are, at best, marginal. The political will is lacking to force users back into public transport, though some attempts are underway (Nottingham Green Partnership, 1995; Cassidy and Flack, 1996). Our lifestyles and personal geographies have become more extensive and complex, so it is difficult to see how these requirements could be met by a return to public transport. But, this is no excuse for many of the changes which have accompanied privatization and deregulation. Minimum service criteria may help overturn some recent undesirable changes, and could be adopted in policy areas which lack political clout (Ghazi, 1995).

For example, season tickets should be available on all forms of public transport, and deserve subsidy if this reduces the number of cars at busy times, or where parking is limited. Cycleways may be encouraged by the Sustrans Project , but even the most extensive recreational cycle network will do little to meet the daily needs of large numbers of existing cycle users (mainly school children), and the potential for much greater cycle use by a large proportion of the population. We should seek the provision for bicycles adopted in the Netherlands, where much of the road system incorporates special cycle provision, and where driver behaviour has been trained to anticipate, and respect, cyclists and pedestrians. This minimum necessary standard would require the retro-fitting of a neighbourhood and sub-neighbourhood scale grid of protected cycleways and 'calmed' street, with links to all schools, employment, shopping and sporting centres and stations, and secure, covered cycle parking at these destinations. To make this case requires overturning the received wisdom that any increase in cycle use raises accident levels, and that bicycles are weather dependent recreational toys, with limited carrying capacity and severe limitations for commuting or 'adult' use (mainly because of their incompatibility with 'smart' office clothing). City greening initiatives, such as the Manchester 2020 Project, offer one way of exploring the potential of more ambitious approaches to 'sustainability', and of realising the scale of change necessary to achieve any real progress (Ravetz and Carter, 1996).

The multi-modal funding packages now required for Transport Policies and Programme bids must be 'cycle-friendly' (Norris, in Baber, 1996). This integrated approach may help overturn limits such as the total Lancashire County Council 1993/94 budget contribution of £49,600 towards the cost of bus shelters (Lancashire County Council, 1994). Commercial deals involving the use of shelters' as advertising hoardings will probably be more important than any increase in public contributions resulting from a more balanced approach to the relationship between different modes of transport. But where will the funding come from for large scale, ongoing commitments of the sort necessary for good quality cycle and pedestrian provision? Analysis of the relationship between different policy areas might be helped by minimum service criteria, even if only to make the case for cross subsidization, and for more effective regulation where falling standards in one policy area have disproportionate effects in another? Decision makers still tend to emphasise the costs of alternatives to the private car, just as the Government has stressed the need to balance the costs of environmental policy against its benefits.

Conclusions

'Pigs might fly'. Why should a decade of deregulation, privatization and declining public services be replace by new, more effective, forms of regulation? Why should politicians and town hall bureaucrats expose themselves to criticism by adopting standards that we can check up on? And why put any faith in the cybernetic possibilities of QA and other forms of self-regulation?

The optimistic reason is that privatization and deregulation are counter-productive. They adversely affect other policy objectives. Re-regulation will reduce external costs (for instance the pollution, congestion and restricted travel patterns associated with poor bus services). More accountable professionals and public servants will be more effective, and may return to the traditions of survey, analysis, plan, so their actions are not limited to the incremental margins, but apply to their whole area of responsibility. The self repairing, professional, nature of QA is particularly attractive, but only in the context of criteria which deal with output. Obsession with bureaucratic process, while potentially useful, easily becomes an excuse for ignoring the real world, and for restricting interest to those few parts that are directly affected by a particular decision or area of administrative responsibility.

As bureaucracy and the professions are as prone to fashion as everybody else, we can expect the language of quality, appraisal and management systems to continue to filter through the complex institutional systems that purport to understand and regulate the built environment, and its social, economic and ecological controls and consequences. This will bring benefits, but these will be constrained by the rigidities and interests which regulate the system of control itself, in particular the conventions and institutions that regulate or own land, water and other crucial resources. It would be a pity if these were permitted to limit the potential benefits from a full appraisal of planning and environmental management. The extension of the 'polluter pays' principle to land, and effective, long term liability for environmental damage will help this process, but it will only become effective once it is accepted that environmental costs must be included in the overall equation.

References

Baber, P. (1996), 'Primary Cycle Routes 'Should Get Priority'', *Planning Week*, 15 February, p. 3.

Barnett, R. (1992), *Improving Higher Education*, The Society for Research into Higher Education and Open University Press: Milton Keynes.

Bennett, S. (1995), 'Environmental Management Systems', *Quality World*, May, pp.324-27.

Bunge, W. (1975), 'Detroit Humanly Viewed: The American Urban Present', Ch 12 in Abler, R., Janelle, D., Philbrick, A. and Sommer, J. (eds.), *Human Geography in a Shrinking World*, Duxbury Press, North Situate: Massachusetts.

Cassidy, S. and Flack, S. (1996), 'Mobility Management in Nottingham: Urban Strategies and Site Plans', unpublished paper, *Annual Conference of the Royal Geographical Society with the Institute of British Geographers*, Strathclyde University, 4th January.

Clark, M. (1988), 'The need for a more critical approach to dockland renewal', in Hoyle, B.S., Pinder, D.A. and Husain, M.S. (eds), *Revitalising the Waterfront: International Dimensions of Dockland Redevelopment*, Belhaven Press: London.

Clark, M., Burall, P. and Roberts, P. (1993), 'A Sustainable Economy', in Blowers, A., *Planning for a Sustainable Environment*, Earthscan Publications: London.

Dauncey, G. (1988), *After the Crash, the Emergence of the Rainbow Economy*, Greenprint: Basingstoke.

Delafons, J. (1995), 'Planning research and the policy process', *Town Planning Review*, Vol. 66, No.1, pp. 83-95.

Department of the Environment, (1994), *Environmental Appraisal of Development Plans*, HMSO: London.

Department of the Environment and Department of Transport, (1994), *PPG 13, Planning Policy Guidance: Transport*, HMSO: London.

Department of the Environment, Welsh Office and National Planning Forum (1994), *Planning Charter Standards*, Department of the Environment: London.

Ellis, R. (1993), 'Quality Assurance for University Teaching: Issues and Approaches', in Ellis, R. (ed), *Quality Assurance for University Teaching*, The Society for Research into Higher Education & Open University Press, Milton Keynes.

Freeman, A.M.III (1982), 'Risk Evaluation in Environmental Regulation', in Magat, W.A. (ed.), *Reform of Environmental Regulation*, Ballinger: Cambridge Massachusetts.

Ghazi, P. (1995), 'Norris Reins in Cowboy Bus Operators', *Observer*, 10th

December, p. 5.

Greed, C. (1993), *Introducing Town Planning*, Longman: Harlow.

Greed, C. (1996), *Implementing Town Planning*, Longman: Harlow.

Green Party, (1996), *Spring Conference 1996, Final Agenda*, Dover, 22-25 February, Built Environment Draft Voting Paper, pp. 33-7.

Hall, D., Hebbert, M. and Lusser, H. (1993), 'The Planning Background', in Blowers, A. (ed.), *Planning for a Sustainable Environment*, Earthscan Publications: London.

Hodgskinson, I. (1995), 'Environmental Management Systems', *Quality World*, February, pp. 94-7.

Houghton, M. (1996), 'Much Ado About Nothing in Planning Performance', *Planning*, No.1186, pp.8-9.

Hughes, D. (1995), 'Environmental Management Systems - An Update', *Quality World*, September, pp. 626-8.

Hutton, W. (1995), *The State We're In*, Jonathan Cape: London.

Keeble, L. (1969), *Principles and Practice of Town and Country Planning*, Estates Gazette: London.

Khakee, A. (1994), 'A Methodology for Assessing Structure Planning Processes', *Environment and Planning B: Planning and Design*, Vol.21, pp. 441-51.

Knox, P. (1982), *Urban Social Geography: An Introduction*, Longman: London.

Knowles, R. (1992), 'The Quality Professional in the 1990s', *Quality Forum*, Vol.18, No.2, pp. 63-7.

Kumar, K. (1995), *From Post-Industrial to Post-Modern Society*, Blackwell: Oxford.

Lancashire County Council (1994), *Public Transport in Lancashire, Policy Review*, LCC: Preston.

Lipietz, A. (1992), *Towards a New Economic Order: Postfordism, Ecology and Democracy*, Polity Press: Cambridge.

Morphet, J. (1995), 'Planning Research and the Policy Process', *Town Planning Review*, Vol.66, No.2, pp.199-206.

Netherwood, A. (1994), 'Applications of Environmental Management Systems', in Fodor, I. and Walker, G. (eds), *Environmental Policy and Practice in Eastern and Western Europe*, Centre for Regional Studies, Hungarian Academy of Sciences: Pecs.

Netherwood, A. (1996), *Environmental Reviews and Environmental Management Systems: Methodologies and Organisational Impacts*, Ph.D. Thesis, Department of Environmental Management, University of Central Lancashire: Preston.

Nottingham Green Partnership, (1995), *Green Commuter Plans, an employer's guide*, pamphlet, Development Department, Nottingham City Council.

Owens, S.E. (1993), 'The Good, the Bad and the Ugly: Dilemmas in Planning for Sustainability', *Town Planning Review*, Vol.64, No.2, pp. iii-vi.

Ravetz, J. and Carter, G. (1996), *Manchester 2020: Sustainable Development in the City Region, Overview*, Town and Country Planning Association, London and CER Research and Consultancy Manchester Metropolitan University: Manchester.

Standing, G. (1992), 'Alternative Routes to Labor Flexibility', in Storper, M. and Scott, A.J., *Pathways to Industrialization and Regional Development*, Routledge: London.

Taylor (1994), 'The Laurie Taylor Guide to Higher Education', *The Times Higher Education Supplement*, Oxford: Butterworth-Heinemann.

Walker, G. and Pratts, D. (1994), 'Environmental Intentions and Environmental Deeds: Analysing the Policy-implementation Divide', in Fodor, I. and Walker, G. (eds), *Environmental Policy and Practice in Eastern and Western Europe*, Centre for Regional Studies, Hungarian Academy of Sciences: Pecs.

Willis, K. (1994), 'Collaborating on collaboration', *Greening Universities*, Vol.1, No.3, October, pp.14.

Wood, C. (1995), *Environmental Impact Assessment: a comparative review*, Longman: Harlow.

14 The rhetoric of Rio and the problem of local sustainability

Bob Evans

Introduction

The 1992 'Earth Summit' in Rio de Janeiro was undoubtedly a major watershed in the development of environmental policy - globally, nationally and locally, and during the five years since Rio, environmental policy, or at least the *vocabulary* of environmental policy has been transformed. The rhetoric of Rio, and in particular the notion of sustainable development inherited from Brundtland (World Commission on Environment and Development, 1987) has been widely promulgated and enthusiastically adopted, particularly in some countries of the North. Furthermore, Local Agenda 21, the local mechanism for pursuing the global goal of sustainability, has an increasing national and international profile.

Britain is one of the countries where LA21 has been adopted with energy and enthusiasm, and this has been due in major part to the work of the Local Government Management Board (LGMB), to pioneering local authorities, and to lobbying of national and international pressure groups, particularly Friends of the Earth and The Worldwide Fund for Nature. According to a survey conducted for the LGMB in February 1996, 91 per cent of UK local authorities are now actively pursing a LA21 programme, and some 40 per cent expected to be able to publish an approved LA21 document by the end of 1996, the target date set at Rio in 1992 (Tuxworth, 1996).

This high level of activity is really quite remarkable, especially when it is remembered that this is a non-statutory process which is being pursued at a time when UK local government is enduring a hostile political and economic climate. As a consequence only limited resources have been available for diversion into LA21 programmes. Furthermore, central government in Britain has offered only limited support, and this only comparatively recently. For example, the government's 'Sustainable Development: the UK Strategy' makes only passing

197

reference to LA21 (HMSO, 1994).

Despite all these constraints the last five years have seen remarkable changes. Two or three years ago, most local government officers and councillors would have no notion of LA21 or the concept of sustainable development, and yet now, superficially at least, these are common currency. It is to the immense credit of UK local authorities, the local authority associations and the environmental pressure groups that this is so.

However, there are some real problems:

- First, there appears to be a substantial and probably unbridgeable 'credibility gap' existing between what might be termed the 'rhetoric of Rio' and the reality of political action at all levels of government.

- Second, the concept of sustainability - or sustainable development - is highly problematic. Confusion, uncertainty and a lack of understanding appear to be endemic to the notion and this inevitably has significant policy consequences.

- Finally, the emphasis placed upon local action is equally difficult. The environmental catechism of thinking globally and acting locally has engendered expectations which may simply not be deliverable at the local level by government or 'local communities'.

Each of these three issues are now considered before making some concluding comments.

The rhetoric of Rio

It is difficult to over estimate the power of the ideas enshrined in the Rio agreements, and specifically in Agenda 21, the agreed global strategy for sustainable development. As many commentators have pointed out, Agenda 21 is a tortuous and difficult to read document. Its 5-700 pages (depending on which version you read) are often opaque and deliberately unclear, imprecise and ambiguous, reflecting the need for compromise in the pre-Summit Prepcoms, and at the Summit itself. Nevertheless, the themes have shone through in Britain at least, due in no small part to the proselytising and interpretation undertaken by the LGMB as part of their LA21 initiatives.

At one level, the rhetoric of Rio is compelling. It is predicated upon the belief that global environmental problems are inextricably interrelated and that they require urgent attention which can only can only be effectively secured through international co-operation. Moreover, although much of this action will need to take place at the international level, for example in the form of agreements concerning CO_2 emissions or the protection of valued environments such as Antarctica, the Rio agreements also recognized the necessity of local action within this global context. It is at this local level that much Rio rhetoric is prominent.

Central to Agenda 21 is the principle that change will have to be 'bottom-up' rather than 'top-down' in character. Governments can encourage and support change, but it is argued that if environmental sustainability is to be a reality, then it will need to be driven by local communities. Thus, the concept of capacity building, the principle of harnessing and enhancing local energies and abilities, is central to the Rio rhetoric, in that it is the central 'means of implementation', of much of the Agenda 21 programme. When this notion is combined with other Rio principles such as partnership, empowerment, participation and consensus, a cocktail of democratization, egalitarianism and subsidiarity emerges which is of central importance to the global sustainability agenda as defined at Rio.

The logic of all this is clear. The sharing of common futures and fates is a pre-requisite for the achievement of global sustainability in the sense that some degree of equity and a greater level of participation in decision making is likely to be needed in order to secure popular consent for difficult policy decisions, nationally and internationally. However, it has to be recognized that these themes, this rhetoric, if taken seriously, could be extremely threatening to established economic and political interests, globally and nationally, and most governments will be reluctant to pursue in any substantial way the policies implied by this rhetoric. In Britain, for example, it is quite clear that central government has no intention of pursuing any of this rhetoric. Despite its centrality to the Rio, decisions all Government reports and statements relating to sustainable development are silent on these matters.

All this is, of course, unsurprising. Although it may be splendid to promote greater political involvement, empowerment and social and economic equity in the pursuit of sustainable development, it is quite clear that those groups possessing political and economic power are highly unlikely to surrender this on the basis of some kind of naive environmental altruism. Nevertheless, at the local level in Britain at least, these ideas have wide currency within local government, environmental groups and most agencies associated with the Local Agenda 21 process.

However, research work currently underway suggests that many local authorities may be consciously ignoring the spirit of much of this rhetoric, although the words themselves may be present in official publications and statements (Evans and Percy, 1996). As a recent LGMB survey has indicated, most UK local authorities have tended to focus their LA21 activities around traditional environmental themes, such as recycling, nature parks or waste management, and have tended to ignore the wider social and economic agendas implicit in the Rio agreements (Tuxworth, 1996). Thus, for example, far from seeking to encourage widespread citizen involvement, or to address the admittedly complex and problematic notions of empowerment and capacity building, most local authorities have tended to 'round up the usual suspects' for a limited consultation exercise, before preparing a traditional report based upon established policy areas and local authority committee structures.

It would, of course, be easy to condemn this as an inadequate and partial response to the inspiring agenda of Rio. However, UK local authorities are clearly operating under a number of substantial constraints. First, there is the question of resources. Most local authorities are conducting their LA21 programmes on a budgetary shoestring and the LGMB survey indicates that most local authorities, some 75 per cent, have simply added LA21 to existing officers' duties (Tuxworth, 1996). Moreover, no financial (and little political) support has been offered by central government.

Second, and perhaps most important, the preparation of LA21 is not a statutory requirement. However, local authorities do have the statutory duty to prepare other plans, particularly land use plans, and it is therefore unsurprising that many have chosen to process LA21 in a somewhat similar manner. LA21 emphasises 'bottom-up' approaches, and is less concerned with the output of a final plan, than with establishing the processes and approaches which will endure into the future. For most local authorities however, accustomed to preparing plan documents which relate to existing professional or departmental responsibilities, for specific and usually statutory ends, this approach has caused some difficulty.

Finally, it seems likely that, despite the stated Rio objective of encouraging hitherto excluded groups (women, young people, black and ethnic minority groups) to participate in the LA21 process, in the main it seems that those groups who have traditionally participated have continued to do so. This means that the educated and articulate property owning middle classes tend to dominate public discourse on these matters. On the limited resources available it its extremely difficult for local authorities to address seriously the problem of 'including' those who have been traditionally excluded for a variety of reasons.

Though some UK local authorities have approached the Rio agenda in an enlightened and progressive manner, there is still some way to go. The majority of authorities are neglecting the social and economic dimensions of LA21, and although the concept of sustainable development is well established, the agenda of empowerment, capacity building and political participation has hardly registered. A significant gap exists between the rhetoric and the reality, which is unlikely to be reduced without significant central government political support and legislation.

Sustainable Development

Sustainable development is now a formal policy goal in the UK at both national and local levels. Moreover, it is also a stated international objective globally, via the United Nations, and at a European level through the Fifth Environmental Action Programme. It is almost impossible to object to the idea of sustainable development, and certainly in Britain, all the major political parties support some variant of the concept centred around the definition made famous by Brundtland. There are however widely varying interpretations, sometimes characterized as light green v dark green, weak sustainability v strong sustainability, or environmentalism v ecologism (Dobson, 1990; Pearce et al., 1990). The variety of interpretations serves to emphasise that sustainability (or sustainable development, depending upon the perspective adopted) is a concept which, superficially at least, has the capacity to span a wide range of environmental positions and social, economic and political interests. It is the very ambiguity of the notion which has made it so popular and so plausible. Indeed it was this ambiguity which facilitated agreement at the Rio Summit between nations with widely differing agendas and perspectives.

It has been argued elsewhere that the character of sustainability is such that, as a policy goal it is intrinsically distinctive and qualitatively different to those policy goals which have been traditionally associated with UK central and local government (Buckingham-Hatfield and Evans, 1996). This has created difficulties for governmental organizations and personnel with which many are still grappling.

The central problem is that the policy goal of sustainability is, in principle, both *long term* and *all-embracing*. The time scale for achieving a sustainable society (if it is indeed achievable) is extremely difficult to specify. However, on current trends it is likely to embrace several generations - at least hundreds of years and perhaps more. Sustainability implies an open-ended commitment to the future, and a responsibility for as yet unborn generations, and as such it represents a policy perspective utterly different from those usually associated with five year electoral cycles.

In addition to the question of time scale, it is extremely difficult to contain questions relating to sustainability to one defined policy area. As has been indicated above, 'environmental' questions, such as pollution, resource usage or recycling rapidly become linked to wider issues of social justice or political participation. Furthermore, given that the environment is not neatly compartmentalized, the 'new environmental agenda' of Rio requires that policy makers operate without regard to traditional professional or departmental boundaries (Blowers, 1993).

There are also problems related to policy achievement. Within traditional policy areas, the objective of policy is usually clear, at least in principle. Thus policy objectives such as 'lower inflation', 'more privatization' or 'more social housing' are apparently quantifiable, and success or failure may be judged in terms of whether particular stated targets have been achieved. In the case of sustainability, however, things are not so clear. Although attempts to identify sustainability indicators look promising, it is unfortunately the case that the complexity of the issues involved make public discourse and political simplification of the issues involved difficult (Pinfield, 1996; Brugmann, 1997).

A further consideration is the temptation to imagine that questions of sustainability are principally scientific or technical in character. Clearly, scientific enquiry has a major role to play in providing the knowledge necessary to understand environmental problems, and to suggest courses of action for policy makers. However, it has to be recognized that the concept of sustainable development is fundamentally political in character, representing as it does, a *belief* in the need for current generations to act as stewards of the earth for future generations. Scientific knowledge can inform policy, but the goal of sustainable development is determined, conditioned and pursued (or neglected) through, and by, the political process.

All of these factors conspire to form a muddy concoction of policy, which many find difficult to understand, and consequently to prioritise. Local politicians in particular may feel that this is all too abstract and removed from the day to day concerns of unemployment, racism, or low pay, and may consign matters of sustainability to the political back-burner. The problem is compounded by the realization that the concept has no agreed definition beyond a nod in the direction of Brundtland, and that all political parties are apparently in favour of it.

One way around this problem may be to regard sustainability in a different light. Sustainability may be represented as what might be termed an 'over-arching societal value', and in this sense it is more akin to notions such as 'freedom', 'justice' or 'democracy', rather than a specific policy goal. Such values are present in most modern societies, and there are likely to be many competing interpretations as to what any particular value might represent, and how it might be achieved. Furthermore, in most instances, there is likely to be a gap between the rhetoric of what any particular value is, or should be, and what might actually exist.

The central point here is that, beyond a superficial representation of sustainability - for instance the Brundtland definition - the concept must be understood as being interpreted politically. The definitions and representations of sustainability, and the agreed policies associated with it will be a reflection of particular social, economic and political positions which have power or ascendancy at any one particular point in time. There is no 'right' definition, and it is futile to search for it. The Brundtland definition is sufficient as a starting point for an appreciation that there are many competing conceptualizations of what sustainability might be, and how we might get there.

The problems of local action

The logic of 'thinking globally, acting locally' is well known and enshrined in Agenda 21, in the European Union's Fifth Environmental Action programme, and in much environmental discourse, and although it is appropriate to recognise the significance and salience of this linkage, it is also necessary to be aware of problems which may be endemic.

In the first place, it is clear that many environmental problems currently facing localities throughout the world are not essentially local in character, however this might be defined. Neither are they likely to be addressed by local action alone. For example, whilst it is clear that local action can make a contribution to climate protection, this is only likely to be effective within the context of international agreements (Collier, 1997). More immediately as far as local environments are concerned, car pollution and traffic congestion may be amenable to solution through local action (schemes for traffic calming, pedestrianization, etc.) but, again these are only likely to have any real bite if they have a national policy, and probably legislative, context.

In Britain, there has been a tendency to emphasise the local action and initiative element of Local Agenda 21. In part this has been an inevitable consequence of the LGMB's campaign to establish LA21 in the face of central government apathy. There can be little doubt that the LGMB has been of pivotal importance in securing such a high profile for LA21 in the UK. However, the need to encourage local authority initiative and innovation has also tended to foster an attitude which seems to absolve central government of any responsibility to participate in the LA21 process, let alone support, co-ordinate or finance it. In Britain, LA21 is a local government show, but although local authorities and local communities clearly have a major role to play, in the absence of national initiative, support and most importantly, legislation, little of substance is likely to occur. Certainly, the massive and long term changes in practice and policy implicit in Agenda 21 are unlikely to be delivered without this legislative imperative.

Local authorities have been placed in an intolerable position. Many, although by no means most, are deeply committed to the ideals of Rio, and are determined to demonstrate that local government can show initiative and take the lead in local environmental policy. On the other hand, it is clear that many of the problems that this policy seeks to address are simply not solvable locally, and in the absence of central government support, the potential for disillusion is high.

In one sense, the rhetoric of Rio itself may be to blame. The emphasis upon local initiative and action, upon capacity building and partnerships with the local community - all this rhetoric has enabled central government in Britain to keep a low profile. Clearly, the rhetoric may have been distasteful to a Conservative government, but this was probably less important than the same government's need to maintain a distance from the overwhelmingly non-Tory local authorities in Britain. It seems as if environmental policy has 'leapfrogged' national government, being driven more by local and international agenda, whether from Rio or Europe.

This lack of national direction and co-ordination is likely to stifle and restrict the pace of innovation. Furthermore, we may begin to move ever more quickly towards a situation where more localities are able to export their environmental problems (waste, cars, pollution, water shortages) to other, poorer, more marginal, peripheral locations. The possibility of increases in spatial environmental inequity is a very real consequence of the Balkanization of environmental initiative and action.

Before leaving this discussion of local action, it is important to reflect briefly on the concept of *local* sustainability, and the extent to which this is a viable or valid tool for policy action. Breheny (1994) argues that a theory of sustainable development, or at least some kind of conceptual framework, is needed as a basis for action, and that such a theory must be of practical use at the local level. Whilst recognising that localities cannot be truly sustainable, he raises the possibility of utilising the concepts of critical local capital and constant local capital as part of a process to minimise resource inputs and waste outputs in localities.

This perspective is useful in that it helps to maintain a focus on the policy process at the local level, and, along with mechanisms such as sustainability indicators, or concepts like ecological footprints, it encourages a reflective policy practice which continually judges policy mechanisms and actions against (sometimes ill-defined) policy goals. However, the central problem remains. Local policy and action can only have a future if it exists within a wider regional or national framework.

Concluding remarks

Sustainability, conceived as a kind of environmental stasis, is a state which will probably never be achieved, and certainly not in the foreseeable future. On the contrary, all the available evidence suggests that the global economy is continuing to move away from sustainability extremely rapidly (Worldwatch Institute, 1997). Furthermore, the current consensus over sustainability is extremely fragile. The oil-producing states and the G7 nations and China, regularly voice disquiet with the concept of sustainable development in international debate, preferring instead 'sustainable economic growth'.

It also has to be recognized that the contemporary debate about environmental sustainability is an exclusive one. Only a comparatively small number of people are involved, and the bulk of the population nationally and globally know little of it and probably care even less. The concept of sustainability is likely to generate policies which run counter to the objective short term interests of a clear majority of most of the populations of developed societies, and many in underdeveloped societies. It asks people to act altruistically, in the interests of as yet unborn children.

Given these not inconsiderable difficulties, can sustainability and the attendant rhetoric of Rio be viewed as a diversion, a slightly naive perspective which will always be sidelined by more pressing political concerns? At one level the answer must be 'yes'. Sustainability as a concept has an enormous capacity to confuse and to exclude, rather than to enlighten and to motivate. For example, recent research has indicated that even those people who are actively involved with the LA21 process may have difficulty in constructing a definition of sustainable development, whilst a concept such as capacity building was understood by virtually nobody (Evans & Percy, 1997). Furthermore, the language of Rio is not easily accessible - 'Local Agenda 21' is enigmatic and cannot be intuitively understood.

Environmental *regulation*, on the other hand may be more explicable and may attract more support. The regulation of pollution, emissions, dumping, development, cars and so on, are policies which can have an immediacy, which may have wider appeal, and which may then lead individuals and non-environmental groups on to wider agenda. If the message of sustainability is to have the wide audience that it needs, there will have to be a greater commitment to education for sustainability - a process of developing an understanding of the scope and magnitude of the environmental problems to be faced, and of the consequences of not taking action. Such a programme of education and raising public awareness will not come cheap, and it is clearly at present outside the capacity of UK local government.

The concepts of capacity building, empowerment, partnership and participation which are central to the rhetoric of Rio, have made little impact on local environmental policy in Britain to date, and the likelihood is that, for the foreseeable future they will continue to remain ideals rather than reality. Without a strong political commitment at a national level to their implementation, with the financial support to develop programmes and the political strength to push them through, it is difficult to see how such radical notions can become common currency in the face of substantial opposition. Moreover, it is not immediately apparent that any UK government would stand to benefit in any way from the implementation of such ideas.

Local action *is* important. Thinking globally and acting locally *does* make sense. The rhetoric of Rio *does* have validity. However, although much has been achieved in Britain during the last few years, there must be serious reservations over what can be delivered in the future without major change. Local government, pressure groups and the local authority associations have put LA21 on the political map. What is needed now is for central government action to build upon this initiative. The local needs to be made national.

Local Agenda 21 needs political support, and the programme needs more central government assistance and guidance. However, important though these are, they are secondary to the pressing need for, firstly a legislative base, and secondly, an adequately funded programme of public education for sustainability.

As the Town and Country Planning Association has so rightly argued, Britain desperately needs a system of environmental planning - an interlinked series of national, regional and local plans which are aimed at securing environmental sustainability (Blowers, 1993). Such plans would need to be holistic and integrated, and would cover a wide range of environmental issues, from pollution, to water resourcing and energy usage, to waste regulation and land use. Environmental plans would need statutory status and would have to be based on democratic discussion and debate. Local Agenda 21 could either be incorporated into such an environmental planning process, or it could co-exist, feeding initiatives into the system. Until local environmental policy and planning has this statutory power and status, LA21 will continue to remain on the margins of both policy-making and public consciousness.

The policy mechanism of environmental planning cannot secure change on its own. The need to motivate and educate for sustainability is the other essential requirement in the move to a more sustainable world. Clearly this cannot be done cheaply, and it will need central government support. However, if the difficult decisions implicit in a move towards a sustainable world are to be achieved , then education for sustainability cannot be seen as an optional extra - it is an essential and indispensable component.

These initiatives alone will probably not deliver the rhetoric of Rio. However, these reforms are a necessary step in the right direction, and will give us a chance to implement some of the essential spirit of Rio.

References

Blowers, A. (ed) (1993), *Planning for a Sustainable Environment*, Earthscan: London.

Breheny, M. (1994), *Defining Sustainable Local Development*, Discussion paper No.23, Department of Geography: University of Reading.

Brugmann, J. (1997), 'Is There a Method in Our Measurement? The Use of Indicators in Local Sustainable Development Planning', *Local Environment*, Vol. 2, No. 1, pp. 59-72.

Buckingham-Hatfield, S. and Evans, B. (1996), 'Achieving Sustainability through Environmental Planning' in Buckingham-Hatfield, S. and Evans, B., *Environmental Planning and Sustainability*, John Wiley: Chichester.

Collier, U. (1996), 'Local Authorities and Climate Protection in the European Union: Putting Subsidiarity Into Practice', *Local Environment*, Vol.2, No. 1, pp.39-57.

Department of the Environment (1996), *Indicators of Sustainable Development for the United Kingdom*, HMSO, London.

Dobson, A. (1990), *Green Political Thought*, Harper Collins: London.

Evans, B. and Percy, S. (1997), *'Has Local Agenda 21 run its course? The prospects for local environmental policy and action in Britain'*, Paper presented at the RGS/IBG Annual Conference, Exeter, January.

HMSO (1994), *Sustainable Development: The UK Strategy*, HMSO: London.

Pearce, D., Barbier, E. and Markandya, A. (1990), *Sustainable Development: Economics and Environment in the Third World*, Edward Elgar: Aldershot.

Pinfield, G. (1996), 'Beyond Sustainability Indicators', *Local Environment*, Vol. 1, No. 2, pp. 151-63.

Tuxworth, B. (1996), 'From Environment to Sustainability: Surveys and Analysis of Local Agenda 21 Process Development in UK Local Authorities', *Local Environment*, Vol.1, No.3, pp.277-97.

World Commission on Environment and Development (1987), *Our Common Future*, Oxford University Press: Oxford.

Worldwatch Institute (1997), *State of the World 1997*, Earthscan, London.

15 Public participation in Local Agenda 21: the usual suspects

Susan Buckingham-Hatfield

Introduction

In 1992, the United Nations Commission on Environment and Development agreed Agenda 21, a programme for global sustainable development. This was a significant achievement since Agenda 21 was debated at the instigation of NGOs. Signatories to this committed themselves to depositing a national plan for sustainable development by 1994 and local areas were required to produce local strategies by the end of 1996. One of the guiding principles of Agenda 21 is that people normally excluded from the decision making process (for example, women, indigenous people and young people) need to be integrally involved in environmental decision making, within a framework which stresses the importance of public participation (United Nations, 1992, Chapter 24). The reason for this inclusive form of participation is that these under-represented groups are seen as having had little impact on the production of environments, although they are sometimes disproportionately affected by them. Therein, however, lies a problem as the structures which traditionally exclude these groups are being invoked to involve them fully. Moreover, greater participation needs a social structure which fosters and encourages this, addressing concepts such as citizenship and empowerment, availability of information and education. The scene is thus set for dissonance between a global agenda heavily influenced by NGO input and its national and local incorporation through political structures.

This chapter will first of all consider the framework in the UK in which public participation in the Agenda 21 process is taking place, paying particular attention to the involvement of women and young people through environmental education. It will review the UK Government's own position on both these areas and will explore the meaning of citizenship in the context of environmental sustainability. From this standpoint, the experience of one area in West London will be reviewed with regard to the response of 'the public' to the process of constructing an agenda for attempting to achieve 'sustainable development'. Whilst a single example is clearly an insufficient basis from which to generalise, Hounslow's approach is similiar to many other Local Agenda 21 processes and it is, therefore, reasonable to suggest that the continued marginalization of women and children is widespread, if not ubiquitous (see, for example, Buckingham-Hatfield and Matthews, 1998, for a commentary on Devon and Australia as well as West London; Evans and Percy, 1998, and Patterson and Theobald, 1998, for other examples of the limitations of participation in London).

The involvement of women

The UK Government response to Agenda 21

Chapter 24 of the United Nations Commission on Environment and Development's Agenda 21 (United Nations, 1992) requires signatories to raise the capacity of women to participate in environmental decision-making and to ensure that structures of decision making facilitate this. This entails a wide programme of literacy, health care, child care, equal opportunities, the ending of discrimination in both the public and the private spheres and the portrayal of women in a positive way. The document recognises that women and children are particularly vulnerable to environmental damage and their perspectives need to be incorporated in environmental decision making by the active and equal participation of women at all levels. Given this unequivocal commitment, it is instructive to consider how the UK government has responded in its submissions to the United Nations outlining how it plans to achieve 'sustainable development'.

There is one mention of women; in the section on non-governmental organizations, the authors claim that '...areas in which NGOs work particularly well include primary health care, the needs of women...' (HMSO, 1994, p.195). Even in sections which discuss population, household formation and income any gendered perspective is missing, as it is in the sections entitled 'Putting Sustainability into Practice' and in the Government's own 'Principles'. To illustrate the distance the UK Government has to travel to come within sight of the United Nations' principles: it is stated in the discussion of fertility that 'The Government believes that couples should make their own decisions about how many children to have...' (ibid., p.8). This presupposes that all parents are part of a 'couple' and marginalises women's rights to determine their own fertility. In his introduction to the document, the Secretary of State for the Environment opens his statement with 'Man has grown used to living as conqueror...' (HMSO, 1994, preface). Even in the more protracted discussions on consumers, there is no acknowledgement that consumption activity may be gendered. By 1996, the government's annual report 'This Common Inheritance' (HMSO, 1996) had dropped this assumption from the introduction, but the neglect of women in the document is still palpable. The only reference is within the international context whereby support is given to improve the health, education, economic, social, political and legal status of women in the South (HMSO, 1996, pp.371-2). The UK government also offers child allowances to enable women from the South to attend training in the UK (ibid., p.373).

Local responses

The Local Government Management Board (LGMB), which is supporting UK local authorities in the development of Local Agenda 21, has published guidelines on including women in LA21. This was one of the last in their series of guidance notes (followed by guidance on involving young people and ethnic minorities), issued towards the end of the period during which plans for sustainable development were being formulated (LGMB, 1996). It assumes that women need to be taught their environmental responsibilities when, arguably, women are most likely to be aware of the environmental implications of their own and other's actions. Women's responses to these actions are, however, frequently constrained through their inability to control decision making due to structural factors (Mellor, 1994, Waring, 1988).

Consistently, research shows that women are considerably more concerned about environmental impacts than are men, not least due to their continual exposure to these issues through their domestic and caring activities (ibid). The author's work in West London (Buckingham-Hatfield, 1994) indicates greater concern and ameliorative action by women compared with men and the Women's Environmental Network (WEN) has shown through its campaigns how powerful and innovative women's action can be in the UK (WEN, 1990, 1991, 1992). These activities are not sufficiently acknowledged in the 'model case studies' highlighted by the LGMB, such as the provision of consumer advice, nappy washing services and the running of organic food co-operatives (LGMB, 1996).

Women, environment, citizenship

Much of the official literature on environmental strategies links action with consumption. This has both positive and negative connotations for the involvement of women in environmental action. By framing women as the main consumers (inasmuch as they undertake the majority of shopping) it can emphasise their power in changing consumption habits. As Ghazi and Jones have recently pointed out in a popular book on 'downshifting': 'Shopping is far from a trivial business. It is...about making political choices. We have a responsibility to get it broadly right. It's your opportunity to change the world and yourself' (Ghazi and Jones, 1996, p.170).

However, there is a danger that women are consigned to this role and can then only express political views through their shopping behaviour. This is increasingly likely in a society which has been recasting 'citizens' as 'consumers'. Despite the recent, much publicised, discussion of citizenship (Andrews, 1991), the rights of the citizen in the UK have arguably been increasingly tied to a person's means to claim and enjoy the rights and responsibilities. Ruth Lister argues that the 'active citizen' is he or she who can 'stand alone, independent before the market, their freedom guaranteed by economic rather than social rights' (quoted in Andrews, 1991, p.13). The exercise of rights and responsibilities also depends on information. In some cases, lack of information may prevent people from playing a full part in civic life, in others, it may be that the information and knowledge that people have to offer is not recognised by decision makers, a point that will be returned to later in this chapter. Elsewhere, Lister (1996) argues that citizenship has a gender dimension, suggesting that, by its emphasis on participation in public life, women are often excluded.

Environmental education and young people

The importance of environmental education has been stressed by the United Nations, through Agenda 21 (1992) and the European Union, through its Fifth Environmental Action programme (CEC, 1992). Both have emphasised the need to educate young people about the environment, placing their faith in younger generations to adopt more environmentally sensitive policies and behaviour. Moreover, both programmes refer to the importance of access to environmental information and to the democratic process in reaching decisions about environmental management. Based on these recommendations, environmental education needs to embrace not only knowledge about environmental systems, but strategies to access and interpret information, assessment of the social implications of environmental impact and use of resources and strategies for intervention in policy making to create change.

Education is only very briefly referred to in the UK response to the United Nations Agenda 21 agreement. This states that schools in the UK are required to deliver environmental education in a 'balanced' way (HMSO, 1994, paragraph 32.18).[1] It continues, 'children should be aware of opportunities that science, technology and other subjects can bring for environmental improvement and not daunted by negative images'..[they] should also be conscious of their responsibilities towards the environment and the contribution that they, as individuals, can make (HMSO, 1994, paragraph 32.13).

This approach arguably constrains what Fien and Trainer value as 'free-thinking' (1993, p19). They suggest that it is a 'fatal mistake to assume that environmental education in schools can make a major contribution to transition to sustainability'(ibid., p.21) and that much more fundamental restructuring is required to achieve sustainability. Consequently, any focus on individual decisions (such as consumption or 'opting out') is insufficient. Nevertheless, it is this stress on the individual and the consumer which pervades the UK documentation on sustainability and which limits the capacity for change which requires a full acknowledegment of social and community rights and responsibilities, in addition to those of the individual, inherent in the concept of citizenship.

Clearly, the approach adopted by the UK Government is designed to ensure that environmental education takes place within the current economic paradigm. The structure of the National Curriculum reinforces this with subject segregation (in direct contrast to the call for environmental education to be inter-disciplinary and holistic, made at the UNESCO conference at Tblisi, UNESCO, 1980) and increased emphasis on competition, career preparation, hierarchies and other structures which, arguably, contribute to environmental problems.

Inter-disciplinarity may be pursued by the cross curricular themes of environmental education, health education, citizenship, economic industrial understanding and careers education and guidance, but these are not mandatory, falling outside statutory requirements. Their delivery, therefore, depends on both the skills and interests of teachers and on the priorities of individual schools (Saunders, Hewitt and MacDonald, 1995). This leads to the second problem which is the capacity of teachers to deliver effective environmental education. Research conducted by Burton in schools in West and South London (1995) suggests that teachers feel ill equipped to teach environmental education. Only 64 per cent of teachers she interviewed admitted that they were able to incorporate aspects of the environment into their teaching. All the teachers interviewed felt that they were inadequately prepared to effectively achieve environmental education in the classroom, although one half had had some kind of training, albeit limited and patchy. The likely outcome of this is for environmental education to be taught as a series of facts and figures 'about' the environment, with some use being made of environmental resources within the school (playgrounds or nature parks) as a vehicle for this learning.

This raises the importance of education outside the classroom. Indeed, Fien and Trainer's (1993) reservations about the marginal value of environmental education in schools is underlined by Burton's findings that school children are more likely to derive their environmental education from TV news (83 per cent) than from school (68 per cent) (Burton, 1995). This raises a serious issue for us all as to the framework in which we 'understand' the environment.

Public participation and local agenda 21: the local dimension in West London

It is through a myriad of influences that our environmental awareness is shaped. We all develop our views through experience which includes education - formal and informal - and exposure to the media. Local knowledge will also be affected by our ability to acquire meaningful information and whether this knowledge is acknowledged may well depend on how we express it. The well documented case of Lois Gibbs' protest against the contamination of Love Canal, USA in 1978, illustrates well how protest is marginalised unless it is formulated in a 'scientific' language (Merchant, 1992). Many commentators on the Local Agenda 21 decision making process have remarked on how there is a wide gap between the language employed by the experts and that by lay people, but especially women (Evans and Percy, 1998, Vallely, 1997, Wickramasinghe, 1998). This section illustrates how 'the public' in an outer West London borough have (or have not) been participants in the Local Agenda 21 process. It raises issues such as privileging points of view, communication and the structural constraints to participation.

Preparing the Local Agenda 21 document

The primary research for the following discussion has been undertaken in the London Borough of Hounslow, through participant observation and documentary analysis. Documentary records of the process in neighbouring boroughs have also been considered. Hounslow encompasses an area affected by major trunk roads and motorways and by Heathrow Airport (see Figure 15.1) and the borough is, at the time of writing, involved in opposing the 5th airport terminal proposed by British Airports Authority. Twenty one and a half percent of the borough's land is designated Green Belt, with a further 13 per cent categorised as Metropolitan Open Land. The River Thames and a number of tributaries offer a great environmental opportunity, but also present environmental problems, for example with a major sewage works serving a large part of Greater London discharging up to 659,000m³ a day into the River Thames.

Hounslow is amongst those boroughs which made an early start on Local Agenda 21. The enthusiasm which launched the process in 1995 arguably prevented a clear analysis of what the genuine participation of women and young people might mean. Consequently the participation process has not significantly challenged previous public participation practice, which Booth (1982, 1996) argues has been markedly unsuccessful in raising the involvement of women. (Typically, she castigates the holding of meetings in the evening, the lack of a creche and remote locations for meetings.)

Figure 15.1 London Borough of Hounslow with major roads and Heathrow Airport indicated

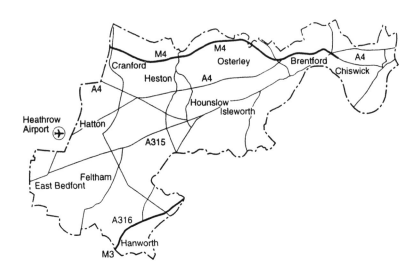

An 'Environmenal Education Day' was held at a local secondary school at which prominent institutions and companies in the area had stalls. There were some lively workshops in which children could act out and model environmental problems. The Environmental Education officer who was hired temporarily to organise the day has resumed teaching and, whilst she and a few other teachers are committed to environmental education, this is by no means widespread in the borough. Apart from this event, there has been no input from young people into Hounslow's proposal for a sustainable environment.[2]

With regard to the involvement of women, no particular considerations were made. The initial launch resulted in the formation, ultimately, of five working groups. In all groups women were outnumbered by men and it was evident that the language element referred to earlier formed a barrier to communication between men (who tended to cast themselves as 'experts', mostly representing organised groups and using the jargon of 'sustainability' and 'futurity') and women (more likely to be representing themselves and noticeably deferring to 'experts' even when their own interests, expressed in the locally immediate issues such as road humps, dog excreta, grafitti and litter, were marginalised).

The group's aim, with the support of the facilitating council officers, was to agree a set of issues and monitoring indicators which were eventually compiled as a draft LA21 document, 3,500 of which were distributed to individuals and organizations. Accompanying this was a questionnaire requesting feedback on the document. The analysis of this feedback (an extremely low response rate of 3.5 per cent) is, however, interesting. One third of responses came from businesses and statutory agencies; 42 per cent of responses were from community or voluntary groups and 23 per cent were from individuals. Only one response was made by a school, which indicates that the document, sent to all schools in the borough, was not used as part of any environmental education process.

Not unexpectedly, twice as many men responded as women: 60 per cent compared to 30 per cent of respondents, 10 per cent did not specify sex. All the businesses and statutory agency responses were signed by men, although women were more predisposed to suggest particular courses of action. This response reflects a gender bias in public participation more generally, and indicates the need to make a more deliberate approach to women if the UN injunction is to be fulfilled.

Analysis

Two factors, I think, have prevented fully participatory discussions taking place on transport and air pollution (the working group which was most closely monitored). One is the organizational structure in which meetings take place in the evening in council offices, which Booth (1982) long ago claimed was destined to exclude large numbers of women from public participation. Such locations have other disadvantages such as the lack of accessibility for residents in other parts of the borough (why not vary the location?). The way in which the venue is organised is also offputting; for example, for a two hour meeting which may entail a total one hour's additional travelling time there are no refreshments laid on (why not provide drinks and light snacks, particularly for those coming straight from work?). The second is the staffing and the control of the dialogue by mostly male, mostly 'expert' 'gatekeepers'. This aspect was most graphically illustrated in a conference held by a neighbouring borough to 'launch' the public participation aspect of LA21. Held in a centrally located adult education college on a Saturday, these organizational strengths were diminished by a list of eight speakers which contained only one woman (a local authority employee) and, in 13 workshops, no women facilitating and only one woman recording.

The specific involvement of women and young people in the LA21 process is, at present, largely unrecorded but the information available (Buckingham-Hatfield and Matthews, 1998) suggests that Hounslow is by no means unique in its failure to involve women and young people significantly in environmental decision making. Such a gender bias in structuring, let alone participating in, the debate is likely to have a significant impact on the extent to which women will enter the sustainability debate and shape future policy. These are research concerns which need to be developed as the process evolves.

Conclusion

Many commentators argue that the real impetus for sustainable development policy in the UK has come from local authorities (Agyeman and Evans, 1994; Buckingham-Hatfield and Evans, 1996), despite a lack of support from central government. This enthusiasm, however, has in many cases, preceeded any in-depth and incisive consideration as to how proper participation may take place. Few authorities, as this chapter illustrates, have employed techniques markedly different from those which have provoked little participation before and, consequently it is difficult to see how the UN invocation centrally to involve previously disadvantaged groups in environmental decision-making is to be materialised and how, therefore, the environment is to become shaped by their needs.

In some respects, despite a long tradition of civic activism and a well developed feminist movement and environmental movement, environmental decision making in the UK is heavily dominated by experts who are predominantly men. National policy declarations, such as 'Sustainable Development: the UK Strategy' (HMSO, 1994) choose not to reflect on the gender dimension and, despite a local government tradition which acknowledges the importance of a women's voice, the Local Agenda 21 process appears to be soliciting public involvement in ways long held to disadvantage women and other groups and individuals ascribed low status. Other strategies which need to run parallel with increased local involvement such as effective environmental education and citizenship building are ambiguously handled and are, anyway, outside the control of local government. Whilst lip service is paid to them (unlike the involvement of women) they are not developed in ways likely to redistribute rights and responsibilities without which a sustainable environment will be unachievable.

Postscript and acknowledgements

The author's position in this research has raised some interesting methodological issues and represents a conflict inherent in applied academic research. My engagement in research into Local Agenda 21 has evolved directly from my strongly held belief that universities have a responsibility to their local community to address, where possible, local needs. As an academic working in the environmental field I both have knowledge which I feel can be useful to local people and policy makers and a personal commitment to doing what I can to improve environmental conditions. Nevertheless, in so participating, I recognise the impact I am likely to have on the outcomes (policy documents, meetings and so on) and have sought to minimise my input unless issues I considered to be important were not being addressed (this was rarely the case). I am also aware that I fall within the category of 'expert' which I have somewhat denigrated in this chapter and do not know how to reconcile, or whether I should even try to reconcile, this with my belief in the importance of 'ordinary' and situated knowledge.

On balance I feel that these tensions are far outweighed by the richness of the material accessed by 'participant observation'. Noting the inflections, body language, minutiae of dialogue and social interaction has been an extremely valuable source of information on the public participation process. I have been lucky in working with commited colleagues who have accompanied me on cold winter evenings to meetings where we could not get anything to eat or drink as well as helping with the questionnaire analysis. I would therefore, like to thank Iris Turner and Kate Theobald at Brunel University as well as Steve McAndrews and his colleagues at the London Borough of Hounslow, fellow Working Group members and participants in the Local Agenda 21 exercise. I would also like to thank Brunel University for the provision of a small research grant which facilitated questionnaire analysis and telephone interviewing.

Notes

1   In the UK, schools are required to deliver a nationally agreed curriculum, under the terms of the 1988 Education Reform Act. Environmental education is primarily delivered through Science, Geography and Technology and pupils are asked to consider the concept of sustainable development and the conflicts of interest which this might involve.

2   Since this chapter was prepared, the London Borough of Hounslow have decided to raise the profile of LA21 in local schools, although it is too early to judge the schools' willingness to be involved.

References

Agyeman, J. and Evans, B. (eds) (1994), *Local Environmental Policies and Strategies*, Longman: London.

Andrews, G. (1991), *Citizenship*, Lawrence Wishart: London.

Booth, C. (1982), Public Participation and Women, in, Polytechnic of Central London (ed.), 'Women and the Planned Environment'. Proceedings of a conference Polytechnic of Central London: London.

Booth, C. (1996), 'Women and Consultation', in Booth, C., Darke, J. and Yeandle, S. (eds), *Changing Places, Women's Lives in the City*, Paul Chapman Press: London.

Buckingham-Hatfield, S. (1994), 'Popular Concerns and the Environmental Agenda: on Involving Women in Formulating Local Responses to Agenda 21', in Fodor, I. and Walker, G. (eds), *Environmental Policy and Practice in Eastern and Western Europe*, Centre for Regional Studies, Hungarian Academy of Sciences: Pecs, Hungary.

Buckingham-Hatfield, S. and Evans, B. (eds) (1996), *Environmental Planning and Sustainability*, John Wiley: Chichester.

Buckingham-Hatfield, S. and Matthews, J. (1998), 'On Including Women: Addressing Gender in Local Agenda 21', in Buckingham-Hatfield, S. and Percy, S. (eds), *Constructing Local Environmental Agendas*, Routledge: London.

Burton, C. (1995), 'Pupils' Awareness of Environmental Issues and Impediments to Implementing Environmental Education in Schools', unpublished dissertation, Brunel University, Department of Geography and Earth Sciences: Isleworth.

Commission of the European Communities (1992), *Towards Sustainability, the Fifth Environmental Action Programme*, CEC: Brussels.

Evans, B. and Percy, S. (1998), 'Has Local Agenda 21 run its course? The prospects for local environmental policy and action in Britain', in Buckingham-Hatfield, S. and Percy, S. (eds), *Constructing Local Environmental Agendas*, Routledge: London.

Fien, J. and Trainer, T. (1993), 'Education for Sustainability', in Fien, J. (ed.), *Environmental Education, A Pathway to Sustainability*, Deakin University Press: Victoria, Australia.

Ghazi, P. and Jones, J. (1996), *Getting a Life*, Hodder and Stoughton: London.

HMSO (1994), *Sustainable Development: the UK Strategy, Cm 2426*, HMSO: London.

HMSO (1996), *This Common Inheritance: UK Annual Report, Cm 3188*, HMSO: London.

Lister, R. (1996), 'Citizenship Engendered', in Taylor, D. (ed.), *Critical Social Policy*, Sage: London.

Local Government Management Board (1996), *Women and Sustainable Development*, LGMB: Luton.

Mellor, M. (1994), *Breaking the Boundaries: Towards A Feminist Green Socialism*, Virago: London.

Merchant, C. (1992), *Radical Ecology, the Search for a Livable World*, Routledge: London.

Patterson, A. and Theobald, K. (1998), 'Local Government Restructuring and Local Agenda 21 in Britain', in Buckingham-Hatfield, S. and Percy, S. (eds), *Constructing Local Environmental Agendas*, Routledge: London.

Saunders, L., Hewitt, D. and MacDonald, A. (1995), *Education for Life; the Cross Curricular Themes in Primary and Secondary Schools*, National Foundation for Educational Research.

United Nations Commission on Environment and Development (1992), *Agenda 21*, UNCED: Geneva.

UNESCO (1980), *United Nations Conference on Environmental Education at Tblisi, USSR*, UNESCO: Paris.

Vallely, B. (1997), Personal communication.

Waring, M. (1989), *Counting for Nothing*, Allen and Unwin: Auckland, New Zealand.

Wickramasinghe, A. (1998), 'Equal Opportunities and Local Initiatives for Environmental Agendas for the 21st Century', in Buckingham-Hatfield, S. and Percy, S. (eds), *Constructing Local Environmental Agendas*, Routledge: London.

Women's Environmental Network (1990), *Disposable Paper and the Environment*, WEN: London.

Women's Environmental Network (1991), *Nappies and the Environment*, WEN: London.

Women's Environmental Network (1992), *Women, Transport and the Environment*, WEN: London.

Printed and bound by CPI Group (UK) Ltd, Croydon, CR0 4YY

21/10/2024

01777088-0001